ENVIRONMENTAL PHYCOLOGY

Climate Change Mitigation and Hazardous Pollutant Cleanup Using Algal Resources

Current Advances in Biodiversity, Conservation and Environmental Sciences

ENVIRONMENTAL PHYCOLOGY

Climate Change Mitigation and Hazardous Pollutant Cleanup Using Algal Resources

Jeyabalan Sangeetha, PhD
Devarajan Thangadurai, PhD
Editors

AAP | APPLE ACADEMIC PRESS

First edition published 2027

Apple Academic Press Inc.
1265 Goldenrod Circle, NE,
Palm Bay, FL 32905 USA

760 Laurentian Drive, Unit 19,
Burlington, ON L7N 0A4, Canada

CRC Press
2385 NW Executive Center Drive,
Suite 320, Boca Raton FL 33431

4 Park Square, Milton Park,
Abingdon, Oxon, OX14 4RN, UK

Apple Academic Press exclusively co-publishes with CRC Press, an imprint of Taylor & Francis Group, LLC

Library and Archives Canada Cataloguing in Publication

CIP data on file with Canada Library and Archives

Library of Congress Cataloging-in-Publication Data

CIP data on file with US Library of Congress

ISBN: 978-1-77964-707-8 (hbk)
ISBN: 978-1-77964-708-5 (ebk)

DOI: 10.1201/9781779647085

About the Book Series: Current Advances in Biodiversity, Conservation and Environmental Sciences

Series Editors:

Jeyabalan Sangeetha, PhD
Assistant Professor, Central University of Kerala, Kasaragod, South India

Devarajan Thangadurai, PhD
Professor and Chairman of Botany, Karnatak University, Dharwad, South India

Nature is that entity composed of three main internally interactive components: biodiversity–human–environment. These are triangularly related aspects of life science, and the interdependency of these three reflects the crucial need of maintenance of equilibrium in their interaction. Both human and biodiversity are two interactive phases of environment that provide a necessary platform for interaction. Being part of biodiversity, human life is almost dependent on biodiversity and its products. This is the point where the human–biodiversity interaction fluctuates due to over-exploitation of biodiversity, where humans take more than the basic needs of human life. Over the past few centuries, this fluctuation in interaction has caused dramatic depletion of biodiversity and thus a drastic change in environment. This has now boomeranged on human life significantly. This variation in the interaction triangle has already travelled the long path of time. Until now, conservationists and life scientists are thinking of restoring the equilibrium in the interaction triangle by taking innovative steps in conserving biodiversity and protecting the environment.

Conservation biology and environmental science are the hotspots of research activities over the past several decades. This has caused much-needed development in observing, identifying, gathering, storing, and analyzing information, social awareness, planning, and advancement in conservation tools and techniques. Conservation biologists have given more importance to building stronger networks of specialists and experts in biology, agriculture, food technology, medicine, mathematics, engineering, architecture, robotics, information technology, and other branches of science and technology to gain innovative ideas to successfully treat the threats to biodiversity and the environment.

Gathering and monitoring information about biodiversity, conservation, and environment is a crucial need of biologists. Surveys from biological survey agencies, establishment of bio-inventories, modern tools and technologies to monitor the biodiversity and collection of data, and strengthening the scientific networks to make awareness, to generate the data, and for the accumulation of both traditional and scientific knowledge about biodiversity conservation and environmental management have been achieved in the recent decades. Conservationists need technological innovations to resolve the threats to biodiversity and environment that are now within reach.

The Current Advances in Biodiversity, Conservation and Environmental Sciences book series publishes titles on a regular basis for a better understanding of biodiversity conservation and environmental management to achieve sustainable development goals for future generations.

Titles in the Series to Date:

Beneficial Microbes for Sustainable Agriculture and Environmental Management
Editors: Jeyabalan Sangeetha, PhD, Devarajan Thangadurai, PhD, and Saher Islam, MPhil

Biodiversity and Conservation: Characterization and Utilization of Plants, Microbes, and Natural Resources for Sustainable Development and Ecosystem Management
Editors: Jeyabalan Sangeetha, PhD, Devarajan Thangadurai, PhD, Hong Ching Goh, PhD, and Saher Islam, MPhil

Algal Genetic Resources: Cosmeceuticals, Nutraceuticals, and Pharmaceuticals from Algae
Editors: Jeyabalan Sangeetha, PhD, and Devarajan Thangadurai, PhD

Mycorrhizal Technology: Managing Plant Stress and Mitigating Climate Change Using Mycorrhizae

Editors: Jeyabalan Sangeetha, PhD,
Abdel Rahman Mohammad Al-Tawaha, PhD, and
Devarajan Thangadurai, PhD

Algal Farming Systems: From Production to Application for a Sustainable Future
Editors: Jeyabalan Sangeetha, PhD, and Devarajan Thangadurai, PhD

Environmental Phycology: Climate Change Mitigation and Hazardous Pollutant Cleanup Using Algal Resources
Editors: Jeyabalan Sangeetha, PhD, and Devarajan Thangadurai, PhD

Data Science and AI for Climate Change: Building a Sustainable Future
Editors: D. Vetrithangam, PhD, Puneet Kumar, PhD, and
Arulmurugan Ramu, PhD

About the Editors

Jeyabalan Sangeetha, PhD
Assistant Professor, Central University of Kerala, Kasaragod, Kerala, India

Jeyabalan Sangeetha, PhD, is an Assistant Professor at the Central University of Kerala, Kasaragod, South India. She has edited and coedited several books in her research areas, which include environmental toxicology, environmental microbiology, environmental biotechnology, bioresources, and environmental nanotechnology. She earned her BSc in Microbiology and her PhD in Environmental Science from Bharathidasan University, Tiruchirappalli, Tamil Nadu, India. She holds an MSc in Environmental Science from Bharathiar University, Coimbatore, Tamil Nadu, India. She is the recipient of a Tamil Nadu Government Scholarship and a Junior Research Fellowship from the University Grants Commission, Government of India, for her doctoral studies. She served as a Dr. D. S. Kothari Postdoctoral Fellow and a UGC Women Postdoctoral Fellow at Karnatak University, Dharwad, South India, from 2012 to 2016, with funding from the University Grants Commission, Government of India, New Delhi.

Devarajan Thangadurai, PhD
Professor and Chairman of Botany, Karnatak University, Dharwad, Karnataka, India

Devarajan Thangadurai, PhD, is a Professor at Karnatak University, South India. He has authored and edited over 30 books with national and international publishers and has visited 24 countries in Asia, Europe, Africa, and the Middle East for academic visits, scientific meetings, and international collaborations. He received his PhD in Botany from Sri Krishnadevaraya University, South India, as a CSIR Senior Research Fellow with funding from the Ministry of Science and Technology, Government of India. He served as a Postdoctoral Fellow at the University of Madeira, Portugal; the University of Delhi, India; and the ICAR–National Research Centre for Banana, India. He is the recipient of the Best Young Scientist

Award with a Gold Medal from Acharya Nagarjuna University, India, and the VGST-SMYSR Young Scientist Award from the Government of Karnataka, Republic of India.

Contents

Contributors

Mrinal Kanti Adak
Department of Chemistry, Indian Institute of Technology Delhi, New Delhi, Delhi, India; Department of Chemistry, Sidho-Kanho-Birsha University, Purulia, West Bengal, India

Deity Anand
Department of Biotechnology, School of Applied Sciences, REVA University, Bangalore, Karnataka, India

Jayaseelan Arun
Center for Waste Management, International Research Center, Sathyabama Institute of Science and Technology, Chennai, Tamil Nadu, India

Namita Bedi
Amity Institute of Biotechnology, Amity University Uttar Pradesh, Noida, Uttar Pradesh, India

Lisianne Brittes Benitez
Environmental Technology Postgraduate Program and Center of Excellence in Oilchemistry and Biotechnology, University of Santa Cruz do Sul, Santa Cruz do Sul, Rio Grande do Sul, Brazil

Dharitri Borgohain
Department of Botany, North Lakhimpur University, Lakhimpur, Assam, India

Gleison de Souza Celente
Environmental Technology Postgraduate Program and Center of Excellence in Oilchemistry and Biotechnology, University of Santa Cruz do Sul, Santa Cruz do Sul, Rio Grande do Sul, Brazil

Gustavo Stolzenberg Colares
Environmental Technology Postgraduate Program, University of Santa Cruz do Sul, Santa Cruz do Sul, Rio Grande do Sul, Brazil

Ajit Das
Department of Chemistry, Sidho-Kanho-Birsha University, Purulia, West Bengal, India

Selvanantham Shanmuganatham Dawn
Center for Waste Management, International Research Center, Sathyabama Institute of Science and Technology, Chennai, Tamil Nadu, India

Sara García-Poza
Department of Life Sciences, MARE–Marine and Environmental Sciences Centre/ARNET–Aquatic Research Network, University of Coimbra, Coimbra, Portugal

Propalika Borpatra Gohain
Department of Botany, North Lakhimpur University, Lakhimpur, Assam, India

Ana Marta Mendes Gonçalves
CFE—Centre for Functional Ecology: Science for People & the Planet, Associate Laboratory TERRA, Department of Life Sciences, University of Coimbra, Coimbra, Portugal; Centre for Functional Ecology – Science for People & the Planet (CFE), Associate Laboratory TERRA, University of Coimbra Campus at Figueira da Foz, Quinta das Olaias, Figueira da Foz, Portugal

Senthilkumar Kandasamy
Department of Food Technology, Kongu Engineering College, Erode, Tamil Nadu, India

Manonmani Kumaraguruparaswami
Department of Food Technology, Rajalakshmi Engineering College, Thandalam, Chennai, Tamil Nadu, India

Silvia Lomartire
CFE—Centre for Functional Ecology: Science for People & the Planet, Associate Laboratory TERRA, Department of Life Sciences, University of Coimbra, Coimbra, Portugal

Enio Leandro Machado
Environmental Technology Postgraduate Program,
University of Santa Cruz do Sul, Santa Cruz do Sul, Rio Grande do Sul, Brazil

Naveenkumar Manickam
Centre for Sustainable Materials and Surface Metamorphosis, Chennai Institute of Technology, Chennai, Tamil Nadu, India

João Carlos Marques
Department of Life Sciences, MARE–Marine and Environmental Sciences Centre/ARNET–Aquatic Research Network, University of Coimbra, Coimbra, Portugal

Prabhakar Mishra
Department of Biotechnology, School of Applied Sciences, REVA University, Bangalore, Karnataka, India

Narasiman Nirmala
Center for Waste Management, International Research Center,
Sathyabama Institute of Science and Technology, Chennai, Tamil Nadu, India

Gislayne Alves Oliveira
Postgraduate Program in Water Resources and Environmental Sanitation,
Federal University of Rio Grande do Sul, Porto Alegre, Rio Grande do Sul, Brazil

Diana Pacheco
Department of Life Sciences, MARE–Marine and Environmental Sciences Centre/ARNET–Aquatic Research Network, University of Coimbra, Coimbra, Portugal

Marcondes Mafaciolli Pacheco
Environmental Technology Postgraduate Program and Center of Excellence in Oilchemistry and Biotechnology, University of Santa Cruz do Sul, Santa Cruz do Sul, Rio Grande do Sul, Brazil

Leonel Pereira
CFE—Centre for Functional Ecology: Science for People & the Planet, Associate Laboratory TERRA, Department of Life Sciences, University of Coimbra, Coimbra, Portugal;
Centre for Functional Ecology – Science for People & the Planet (CFE), Associate Laboratory TERRA, University of Coimbra Campus at Figueira da Foz, Quinta das Olaias, Figueira da Foz, Portugal

Tiele Medianeira Rizzetti
Environmental Technology Postgraduate Program and Center of Excellence in Oilchemistry and Biotechnology, University of Santa Cruz do Sul, Santa Cruz do Sul, Rio Grande do Sul, Brazil

Lúcia Helena Ribeiro Rodrigues
Postgraduate Program in Water Resources and Environmental Sanitation,
Federal University of Rio Grande do Sul, Porto Alegre, Rio Grande do Sul, Brazil

Samraj Sadhappa
Department of Petroleum Engineering, JCT College of Engineering and Technology, Coimbatore, Tamil Nadu, India

Andrea Marisa Sanchez-Barrios
Environmental Technology Postgraduate Program and Center of Excellence in Oilchemistry and Biotechnology, University of Santa Cruz do Sul, Santa Cruz do Sul, Rio Grande do Sul, Brazil

Rosana de Cassia de Souza Schneider
Environmental Technology Postgraduate Program and Center of Excellence in Oilchemistry and Biotechnology, University of Santa Cruz do Sul, Santa Cruz do Sul, Rio Grande do Sul, Brazil

Padmashree Sekar
Department of Biotechnology, School of Applied Sciences, REVA University, Bangalore, Karnataka, India

Ashutosh Srivastava
Department of Zoology, Siddharth University, Kapilvastu, Siddharth Nagar, Uttar Pradesh, India

Dhaval Kumar Shrivastava
Amity Institute of Biotechnology, Amity University Uttar Pradesh, Noida, Uttar Pradesh, India

Maiara Priscilla de Souza
Environmental Technology Postgraduate Program and Center of Excellence in Oilchemistry and Biotechnology, University of Santa Cruz do Sul, Santa Cruz do Sul, Rio Grande do Sul, Brazil

Munusamy Subathra
Department of Biotechnology, Aarupadai Veedu Institute of Technology, Vinayaka Mission's Research Foundation, Paiyanoor, Tamil Nadu, India

Bhaben Tanti
Department of Botany, Gauhati University, Guwahati, Assam, India

Burcu Ertit TaŞtan
Vocational School of Health Services, Gazi University, Ankara, Türkiye; Life Sciences Application and Research Center, Gazi University, Ankara, Türkiye

Basheer Thazeem
Post Graduate and Research Department of Zoology, Maharaja's College (Government Autonomous), Ernakulam, Kerala, India

Abbreviations

ACC	algal carbon conversion
APAH	alkylated polycyclic aromatic hydrocarbon
BIQ	bio-intelligent quotient
BOD	biochemical oxygen demand
BPA	bisphenol A
CA	carbonic anhydrase
CCM	CO_2 concentrating mechanism
CCS	carbon capture and storage
CCU	carbon capture and utilization
CDR	carbon dioxide removal
CO_2	carbon dioxide
COA	coenzyme A
COD	chemical oxygen demand
DON	dissolved organic nitrogen
EDC	endocrine-disrupting chemical
EPS	exopolysaccharides
HAB	harmful algal bloom
HOC	hydrophobic organic compound
HRAP	high-rate algae pond
HRT	hydraulic retention time
IPCC	Intergovernmental Panel on Climate Change
LLGHGs	long-lived greenhouse gases
MABA	microalgal-bacterial aggregate
MEM	marine ecosystem model
NRE	nutrient removal efficiency
OA	ocean acidification
PBR	photobioreactor
POC	particulate organic carbon
PSP	paralytic shellfish poisoning
RNS	reactive nitrogen species

ROS	reactive oxygen species
SRM	solar radiation management
WWT	wastewater treatment

Preface

Environmental phycology refers to the biological and biotechnological applications of algae for various environmental applications such as pollution control, wastewater treatment, biofuel production, and heavy metal remediation. Environmental phycology employs both microalgae and macroalgae for environmental applications. In agriculture, microalgae are used for soil reclamation. Since microalgae have the ability to fix atmospheric nitrogen, they are used as biofertilizers to increase soil fertility. Microalgae are also known to improve the physicochemical properties of soil, such as pH, electrical conductivity, potassium, nitrogen, and phosphorus. Species of *Spirulina, Anabaena, Nostoc, Gloeotrichia, and Aulosira* are commonly used as biofertilizers because of their high efficiency in nitrogen fixation. Algal biomass is considered a very good source of energy because it is easily available and has good carbon dioxide fixation capability. Biochar is produced from algal biomass using a slow pyrolysis process to form a product that is rich in carbon. This can be used to improve the pH of acidic soils. In addition, algal biochar is used as an adsorbent in the wastewater treatment process since it contains inorganic elements and several functional groups.

The presence of certain functional groups in microalgae makes them good adsorbents for the removal of organic molecules in wastewater. Bioethanol, biohydrogen, and biodiesel are synthesized from cyanobacteria and photosynthetic algae. Because of their capacity to store large quantities of lipids in their cells and their high photosynthetic ability, algae are considered a good source for biodiesel production. *Chlorella vulgaris* is used for the production of biofuel and biomass in significant quantities. *Cladophora fracta, Chlorella protothecoides, Botryococcus braunii, Gloeocapsa alpicola, Spirulina platensis, Chlamydomonas reinhardtii, Platymonas subcordiformis, and Ulva lactuca* are some other promising algal species that are used for biofuel production worldwide. Microwave irradiation–assisted transesterification technology requires less time and less heat energy to enhance the production of biodiesel from algal biomass.

Algal species such as *Anabaena, Spirulina, Oscillatoria, and Chlorella* are used for the treatment of acid mine drainage contaminated with heavy metals such as aluminum, iron, lead, and tin. These metals are present in large quantities, and more time is required for their removal because they can have a negative impact on the environment. Hence, the removal of these metals using phycotechnology is considered eco-friendly and economical. Algae have the ability to sequester heavy metals because of their biosorption properties. The metal oxides can be recovered by drying the algal mass, and biofuel can also be produced from the stored algal mass. Hydrothermal liquefaction is a recently developed technique used to recover metals present in the algal biomass and to utilize the resulting biomass for biofuel production.

Algae wheel is another technology used for the treatment of wastewater. In this technology, algal biofilm is integrated with existing biological wastewater treatment plants. This technique uses the symbiotic relationship between bacteria and algae. The photosynthetic ability of thickly grown algal mats on the wheels used for rotating water in wastewater treatment plants is utilized for treating wastewater. The oxygen produced as a result of photosynthesis during this process is used by the bacteria for improved performance of the degradation process. This technique is very effective in water pollution control and wastewater treatment because it is simple, requires low energy, is cost-effective, and produces no odor or noise.

The present book is a unique book that deals with different approaches for the mitigation of climate change and environmental cleanup using micro- and macroalgae. This book specifically addresses the management of environmental pollution using microalgae for a sustainable environment and carbon sequestration. The book mainly covers the topics of biofuel production, waste treatment, algal remediation, carbon sequestration, and climate stabilization using algae and their biomass.

—Editors

CHAPTER 1

Algae-Mediated Environmental Sustainability Solutions for Climate Change

SAMRAJ SADHAPPA,[1,*] SENTHILKUMAR KANDASAMY,[2,*] NAVEENKUMAR MANICKAM,[3] MANONMANI KUMARAGURUPARASWAMI,[4] and BASHEER THAZEEM[5]

[1]*Department of Petroleum Engineering, JCT College of Engineering and Technology, Coimbatore, Tamil Nadu, India*

[2]*Department of Food Technology, Kongu Engineering College, Erode, Tamil Nadu, India*

[3]*Centre for Sustainable Materials and Surface Metamorphosis, Chennai Institute of Technology, Chennai, Tamil Nadu, India*

[4]*Department of Food Technology, Rajalakshmi Engineering College, Thandalam, Chennai, Tamil Nadu, India*

[5]*Post Graduate and Research Department of Zoology, Maharaja's College (Government Autonomous), Ernakulam, Kerala, India*

*Corresponding author

ABSTRACT

Algae are progressively critical in considering the current and long-lasting impacts of environmental change, where green growth management seeks to maintain carbon stocks to provide a source of renewable energy. Adjustment activities that work to reduce a system's vulnerability in an evolving atmosphere can help support beneficial carbon outcomes. Algae

Environmental Phycology: Climate Change Mitigation and Hazardous Pollutant Cleanup Using Algal Resources. Jeyabalan Sangeetha and Devarajan Thangadurai (Eds.)

DOI: 10.1201/9781779647085-1

are among the most significant organisms on Earth, as they provide almost all of the Earth's oxygen, which we need to breathe. Fuels used in vehicles and planes might be derived from algae in the future. The main objectives of this chapter are to discuss algae as an alternative solution for climate change, its effects on carbon, the relationship between algae and climate change, the benefits of algae for climate change, and how governments consider the risks to their communities to improve health care services.

1.1 INTRODUCTION

Algae are photosynthetic organisms that convert carbon dioxide (CO_2), light, and water into dense nutrients. They are also used as food for existing living things. Algae can produce biomass 10 times more swiftly compared to terrestrial plants, and less than 10% of the total area is required to generate a comparable quantity of biomass (Brennan and Owende, 2010). They can cultivate on unproductive land and do not compete with other crops for land due to their low water requirements. They can be managed more resourcefully compared to land crops. Algae can help avoid the rigorous water practices and eutrophication related to current farming.

Algae provide a valuable source of carbon-neutral fuels, aqua fuels for the farming and aquaculture sectors, and high-protein food products for human consumption (Andersen, 2014). The carbon-neutral perspective, which does not discharge CO_2 into the atmosphere, has evolved. Most light vehicle fleets are expected to be electrified by 2030–2050, and there will be less need for fuel; however, there may still be a requirement for jet aviation and transport in the foreseeable future. About 10–15 years remain to address global climate change and food safety crises, as it may take 2 to 3 decades to realize solutions at the required scale to resolve these worldwide issues (Chisti, 2008). The main objectives of this chapter are to discuss algae as an alternative solution for climate change, their effects on carbon, the relationship between algae and climate change, the benefits of algae for climate change, and to address how governments consider the risks to their communities in improving health care services.

1.2 ROLE OF ALGAE IN THE ENVIRONMENT

- Algae synthesize oxygen in the environment.
- Algae support aquatic ecosystems by providing energy for the food web.

- Algae produce sugar from H_2O and CO_2 via photosynthesis.
- Algae can reduce operational and capital costs and produce algal fuels such as bioethanol, biodiesel, biomethanol, and so forth.
- Algae used for wastewater treatment (WWT) have more advantages than conventional WWT.
- Algal bioreactors are utilized for the manufacture of biofuels.
- Algae can be used to trap fertilizers present in runoff.

1.3 CLIMATE CHANGE EFFECTS ON CARBON

Figure 1.1 demonstrates climate change, which is affecting the environment worldwide. Numerous changes are also likely to continue in the future. Opportunities to mitigate atmospheric greenhouse gas emissions (Chung et al., 2013) are generating interest in controlling carbon within ecosystems, highlighting the sequestration of CO_2 and providing a basis for renewable energy. However, simultaneously, modifications in the Earth's climate system are impacting forests in numerous ways, which could affect the emissions of carbon and other GHGs. New climate conditions may

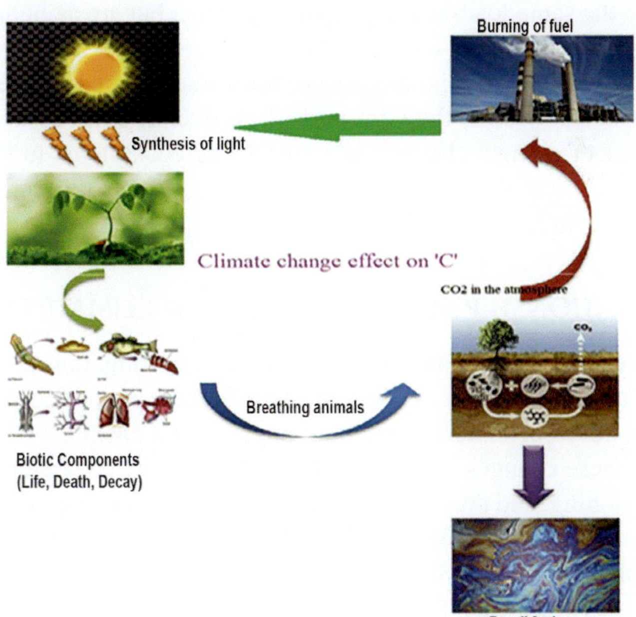

FIGURE 1.1 Flow diagram showing the effect of carbon on climate change.

shift faster in some locations than species can disperse, creating uncertainty about the future vegetation composition of these new habitats.

Weather events and climate conditions, rather than gradual changes in temperature or precipitation, are increasing climate extremes, and as forest ecosystems change, water produced from forest land may become more variable and of lower quality (Davis et al., 2003). Climate change impacts in forest environments are likely to create losses of ecological units in various areas.

1.4 MANAGEMENT OPTIONS

Algae can convert CO_2 from air power plants or steel processing exhaust into biomass, which can then be used to create important carbon fibers. These fibers are similar to conventional strands and can subsequently be utilized in various existing processes. Another significant field of use would be the construction industry, which (Singh and Dhar, 2019) represents a critical extent of global CO_2 emissions. Carbon fibers can replace steel in construction materials; their strength can save concrete, and stone reinforced with carbon fiber can even be used to produce beams that have the same load-bearing capacity as steel but are as lightweight as aluminum.

Adaptations and mitigations are not substitutes for one another (Melillo et al., 2014); however, they are part of an overall strategy to minimize the impact of climate change. Adapting algae ecosystems to changes in climate is critical for maintaining and reducing losses of carbon to the entire atmosphere.

1.5 RELATIONSHIP BETWEEN ALGAE AND CLIMATE CHANGE

Oceans retain both heat and carbon from the air, mitigating the effects of global warming in nature. However, in the last couple of decades, human activities have been altering essential physical and chemical properties of the sea. These changes are expected to affect marine ecosystem structure and functioning, and thus the potential to modify the cycling of carbon and nutrients on the sea surface, with likely feedback on the climate system (Janowiak et al., 2017; Melillo et al., 2014).

Increases in seawater temperature will reduce CO_2 solubility, decreasing anthropogenic carbon uptake by 9–15%. Global warming will also intensify the hydrological cycle, increasing precipitation in some areas and evaporation in others. It is expected that higher temperatures will promote photosynthetic activity, enhancing primary productivity in areas with no nutrient limitation, but simultaneously, higher temperatures will promote heterotrophic metabolism, intensifying top-down control of phytoplankton assemblages and reducing primary productivity (Notoya, 2010).

Understanding how stressors influence natural networks and ecosystems and the prevalence of synergistic or non-synergistic interactions remains a challenge for scientists. Many of our current expectations from investigations of the effects of single climate stressors on the productivity of algal communities under future ecological conditions will likely need reconsideration after the widespread acknowledgment of interactions and feedbacks among physical and biological factors (Olguín, 2003). An additional challenge is to upscale results from studies to significant global impacts. The primary productivity of the ocean, on the contrary, results in the world being described in recent literature as negative in some studies and positive in others, reflecting our great uncertainty regarding climate change and its consequences.

1.6 CHALLENGES FOR CLIMATE–CHANGE SOLUTION

Difficulties with forestation as an atmosphere-change arrangement incorporate the relocation of land utilized for cultivating, logical and innovative troubles in estimations and observing, and restricted open financing for carbon gainful administration, as indicated by the World Resources Institute (WRI). Planting as many trees as could be expected under the circumstances as fast as conceivable reduces 9 gigatons of carbon a year, but it would likewise increase food costs by 80% by 2050. Algae can be used in various manners to decrease carbon in the environment; however, the proficient answer for putting away CO_2 can be effectively utilized in an assortment of other sustainable and industrial items or materials, from sneakers to steel alternatives.

1.6.1 ALGAE AS CARBON SEQUESTRATION

Plants and algae naturally sequester CO_2, as shown in Figure 1.2. Algae follow a similar process but assimilate carbon into additional algal biomass. Currently, numerous physicochemical carbon capture and sequestration systems are combined and categorized as carbon capture and storage (CCS) procedures. CCS operates through three major storage processes: CO_2 capture, CO_2 transportation, and CO_2 storage. CO_2 capture is carried out from large point sources, such as power plants and cement manufacturing plants. The separation and capture of CO_2 from other flue gas components are generally performed using the following strategies: (i) chemical absorption; (ii) physical absorption; (iii) membrane separation; and (iv) cryogenic distillation. Algae can consume more CO_2 than trees because they can cover a greater surface area, grow faster, and be more easily controlled in bioreactors.

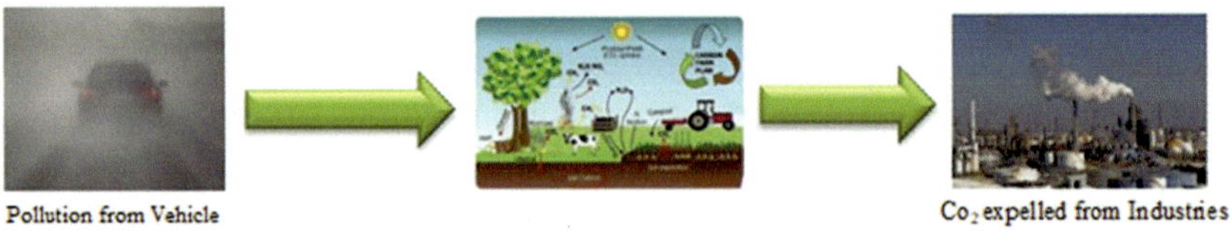

FIGURE 1.2 Flow diagram for carbon sequestration.

1.6.2 ALGAE AS FOOD

Algae are the most significant organisms on earth, as they provide almost all of the earth's oxygen, which we need to breathe. Fuels used in vehicles and planes might be fueled by algae in the future. Algae assimilate CO_2 as they grow rapidly and yield diversely. Algae's high biomass yield and capacity to grow on non-arable land, therefore avoiding competition with food crops, mean that algal food production can help reduce land-use change emissions through the balance of conventional agricultural growth. Additionally, saltwater algae for food items may lead to significant freshwater savings; however, nutrients for algae growth should come from waste streams to ensure sustainability, by reducing the land demand of food production while simultaneously enhancing food and energy security. Algae can further enable the development of terrestrial bioenergy technologies, including those using CCS.

1.6.3 ALGAE AS MATERIAL

It is probably the luminous, translucent 3D-printed vessels by studio Marenbeek and Dros, self-described as designers of the unusual in a quest to fuel an alternative to fuel and oil-based synthetics. The studio has made bioplastic from food waste streams, including cacao and potato; algae will help to be used as both a filler and a pigment for a biopolymer that can be used in conventional 3D printers (Vose et al., 2012). One potential future for algae is the scaling of the industry to pose a new set of energy costs that small-scale production can keep low. Based on the current processes used by designers working with the material, what seems more likely is that algae becomes increasingly popular for a nimble quality that sets it apart. A source of pollution that contributes to global warming can be transformed into a way to deduct CO_2 from the environment by using algae as a raw material.

1.6.4 ALGAE AS BIO-ENERGY

The chance of global climate change is essentially attributed to anthropogenic greenhouse gas emissions from fossil fuel use, deforestation, and expanded development of farmland; however, stochastic mega-events, such as volcanic eruptions, cannot be discounted as principal contributors to fuel emissions. It is a reality that oil runs the modern world, and there is an entrenched hydrocarbon economy, which has been the motive of more than one conflict in oil-producing regions (Oncel, 2013). Oil and its products are vital as they provide low-cost fuel for the internal combustion engine that is required for transportation. In addition, diverse grades of oil are produced for numerous applications, with wide-ranging uses from candle wax to plastics to much-needed medicines (including aspirin, psychotropic pills, and antihistamines). An assessment of the restrictions and geopolitical disruption of world reserves, along with the perceptions of declining availability and increased costs, has led to increasing efforts to develop resources of alternative, renewable energies. Among these are biofuels, including those made from algal-derived feedstocks. The need for energy production (bio-energy, solar energy, wind energy, etc.) partnered with environmental conservation processes, without neglecting the global CO_2 sink, is needed.

Microalgae offer several advantages compared to land plant life because of their excessive rate of growth, capacity to grow on non-arable land, and their considerable content of strength-rich compounds that may be optimized under controlled situations, harvested, and chemically transformed into fuels. They can probably produce biofuels in several ways, including biodiesel, biogas, bio-hydrogen, and bioethanol, and it has been proven that they are greener than terrestrial plants in converting sunlight to biochemical power. Indeed, a few species have a fast rate of oil productivity. To fully compete with fossil fuels, principal challenges remain, including the poor returns on both economic and energy investments, which weakens the argument for the feasibility of business scale-up and, subsequently, sustainability. Economic viability is an important pillar of sustainability. The organic truth remains firm that algae may be used as biomass for many types of biofuels. This has been proven to be possible for a long time, but despite many contemporary advances, the unsubsidized scale-up from pilot plants to long-term viability of industrial operations remains elusive. However, it will likely be instructive to look at the numerous operations presently advancing in these endeavors. Unfortunately, decreased oil elements caused by natural disasters and/or future global conflicts and wars can be the important economic situations that tip the balance in favor of economically viable biofuels. Of course, the demand for these biofuels is maintained by the attachment to modern energy sources, but developments of different energy resources, such as hydrogen, for instance, may reduce the demand for "expensive" photosynthetically derived options (Field et al., 2012). Today, international attention within societies has been kept alive with recurrent news items and promotional fancy slogans about biofuels. However, there is some speculation that the political jargon of the "future" of microalgal biofuels is meant only to assuage certain factions interested in alternative energy sources. The announcement of the need for a fourth-generation biofuel from genetically modified algae to enhance the production and commercialization of fuel oil should rekindle public interest, but since GMO and photosynthetic life bureaucracy could be involved, there is a significant potential for essential conflicts ahead of that initiative. Genetic modification remains a touchy issue, and more controversial debate may arise, requiring increased efforts of communication at the edge between technological concepts and social order. One of the limitations restricting the development of microalgal biofuel technology lies in the fundamental limitations on the physico-chemical efficiencies of light harvesting of the biomass before the inefficiencies of mass cultivation are considered.

1.6.5 REMEDY FOR ENVIRONMENT

The bioremediation properties of algae (micro and macro) remain an untapped remedial capability for pollutants from various sources, including the biosorption of heavy metals from industries. Phycoremediation is the use of the biotransformation ability of algae for the removal of pollution and pollutants from wastewater and CO_2 from effluent air streams (i.e., fuel and cement industries, among others). On the other hand, the technologies being "reinvented" and refined are not new concepts; for example, the concept of high rate algal ponds (HRAP) was developed by Oswald et al. (1957) in the mid-1950s, and since its inception, the technology has spread worldwide. It is currently utilized for commercial algal production and/or for the development of phycoremediation technology for domestic or agricultural wastewater streams, including those found on dairy farms, piggeries, among others. Currently, the use of WWT HRAPs for algal production and biofuel conversion has received little attention (Adams, 2019). The demonstration of promising performance by microalgae in a five-hectare-scale wastewater HRAP over seasonal scales in a temperate climate appears to be an economically viable way for biofuel production to occur with minimal environmental effects.

The treatment of domestic sewage and its conversion to energy have recently been highlighted in the media with the completion of experimental architectural initiatives integrating algae within the glass façades of buildings, including the BIQ (bio-intelligent quotient) building in Hamburg. These bioreactor façades are expected to provide services such as the production of energy and biomass, sound insulation, and heat insulation.

Recent hot topics include the use of algae for "carbon capture," with the biomass used for the extraction of bioactive compounds and agricultural inputs, and bioenergy as a potential "derivative." For example, the aim of the National Research Council's (NRC) Algal Carbon Conversion (ACC) Flagship is to provide Canadian industry with a cost-competitive, revenue-generating strategy to divert CO_2 emissions into algal biomass (Rhodes, 2010), which could then be processed into biofuels and other marketable products. This initiative is presented as an opportunity to address environmental, energy, and economic issues associated with commercial emissions. It explores how CO_2 emissions can be sustainably and profitably transformed by algae into valuable products without displacing food crop sources. CO_2 sequestration through marine macroalgae has been considered in the last decade as a potential sink for anthropogenic CO_2 emissions, especially supported within the "Asian Network for Using Algae as a CO_2 Sink." Through experiments conducted in South Korea since 2006, seaweed cultivation may be one measure, among others, for mitigating global warming through CO_2 capture and oxygen production via algal photosynthesis. These initiatives might appear to promise environmental services and even opportunities for emissions trading within the context of the Kyoto Protocol. Further economic evaluation and integrated project business plans are needed to assess the value, as well as additional analysis to measure the effective CO_2 scale, particularly in short-lived algae.

1.6.6 DISPLACEMENT OF LAND USED IN FARMING

Land use is evaluated as the surface area required to produce 1 kg of algal biomass for more than one year. Climatic conditions influence land use by affecting profitability, and land use is higher for closed photobioreactors (PBRs) than for open ponds under all atmospheric conditions considered because of differences in anticipated system productivities (Oswald et al., 1957). The estimated land use for PBRs strongly depends on reactor density, which was hypothetically fixed at one reactor per m^2 in this case study. Increasing reactor density should increase areal productivity until algae assimilate all the sunlight reaching a square meter of flat surface (Park et al., 2011). The distance between PBRs can be optimized to maximize productivity, and future optimization studies should be used to refine the land use assessment of full-scale algal development in PBRs.

1.6.7 SCIENTIFIC AND TECHNOLOGICAL DIFFICULTIES IN MEASUREMENTS AND MONITORING

Algae are a critical part of natural checking programs for evaluating water quality; they are qualified to water quality appraisal as a result of their supplement necessities, quick generation rate, and short life cycle. Algae are significant pointers of condition circumstance since they react quickly to both subjective and quantitative creation of species in a wide scope of water circumstances due to adjusting in water science (Jacquin et al., 2014), for example, increments in water contamination dependent on local squander and balance the organization of genera that can endure these circumstances.

From an environmental and public welfare point of view, large quantities of nutrients and phosphorus that flow into natural reservoirs such as lakes and other marine ecosystems cause eutrophication. The nitrogen (N):phosphorus (P) ratio determines which algal genera are dominantly present or absent in these nutrient-impacted aquatic ecosystems (Arnold et al., 2018). The following are the important properties of algae, and these properties make algae more suitable for biomonitoring:

1. Algae are autotrophic and are positioned at the interface between the environment and the biotic components of the food web.
2. Green algae are fundamentally sessile; they cannot move to avoid contamination and therefore may disappear.
3. All algae have short life cycles and thus exhibit a rapid response to change, while the community persists long enough to integrate interactions over time.
4. Algae are smaller in size than other biota and are potentially more sensitive to contamination at lower concentrations.

1.6.8 HUMAN HEALTH AND PRODUCTIVITY

If the temperature raises 4.5°C by 2090, 9300 more people per year may die in America because of the increasing global temperature. The yearly losses related through excessive temperature, and related fatalities alone are proposed to be $14 billion. Ever-rising warmness and precipitation may also put in to the hazard of waterborne and foodborne illness and allergies. The precipitation of insects may outspread diseases like zika, dengue and Lyme disease into novel territory (Arnold et al., 2018). Extreme weather and climate-related natural disaster can also aggravate psychological healthiness related issues. The most susceptible people and poor people may mostly be influenced by these health-related impacts.

1.6.9 CLIMATE CHANGE AND OPPORTUNITY

There will be greater opportunities in clean energy, resilient and green buildings, and energy efficiency. Hybrid and electric vehicle production in the public transportation sector is expected to grow. Carbon capture and sequestration, as well as the use of captured CO_2, present opportunities. As Arctic ice melts, new transportation routes will open for trade, significantly reducing travel time. The warming Arctic could also offer additional opportunities for oil and gas drilling. Climate satellites and radar technology will be in demand to monitor extreme weather (Sutherland et al., 2013).

1.7 BENEFITS OF ALGAE FOR CLIMATE CHANGE

Algae's battle against climate change is that it not only removes carbon from the air but is also the most efficient at storing CO_2 and can be easily used in sustainable industry products. Unlike trees, algae can do this without taking up large amounts of land and freshwater, which fuel natural forests and agriculture.

When an alga absorbs CO_2 from the atmosphere, power plants, and steel processing emissions, it can be converted into algae oil. This can be used to produce carbon fibers economically (Stephenson et al., 2011). These fibers can then be used to create both lightweight and high-quality materials that are carbon neutral.

1.8 GOVERNMENT

Government ministries should carefully consider the risks and hazards their populations face before a natural disaster occurs. They should invest in resilient measures such as improving infrastructure, restoring aquatic resources, minimizing unemployment, and engaging communities in essential systems, including enhancing health care–related services. Government officials are currently assessing the Green New Deal in relation to climate change mitigation.

1.9 CONCLUSION

Algae are an increasingly viable option for addressing many costly climate change challenges. Next-generation industries are recognizing this opportunity and investing in it; however, it is also something that individuals can consider as they seek to reduce their impact on the environment. Global challenges can be addressed through two approaches: collective action and human-centered technology that improves outdated methods. Everyone can be part of this change and support organizations that are leading the way. We must consider alternative options that are environmentally sustainable, not just cost-effective. Greater investable capital is needed for long-term solutions that address significant issues and create employment opportunities; investment options should allow for more stable returns and long-term planning, strategic thinking, and implementation.

KEYWORDS

- **algae ecosystems**
- **carbon fibers**
- **forest environments**
- **hydrological cycle**
- **marine ecosystems**
- **phycoremediation**

REFERENCES

Adams, B. (2019). *Green Development: Environment and Sustainability in a Developing World*. Routledge.

Andersen, R. A., & Bux, F. (2014). *Biotechnological Applications of Microalgae: Biodiesel and Value-Added Products*. CRC Press.

Arnold, U., Brück, T., De Palmenaer, A., & Kuse, K. (2018). Carbon capture and sustainable utilization by algal polyacrylonitrile fiber production: Process design, techno-economic analysis, and climate related aspects. *Industrial and Engineering Chemistry Research, 57*(23), 7922–7933.

Arnold, U., De Palmenaer, A., Brück, T., & Kuse, K. (2018). Energy-efficient carbon fiber production with concentrated solar power: Process design and techno-economic analysis. *Industrial and Engineering Chemistry Research, 57*(23), 7934–7945.

Brennan, L., & Owende, P. (2010). Biofuels from microalgae—A review of technologies for production, processing, and extractions of biofuels and co-products. *Renewable and Sustainable Energy Reviews, 14*(2), 557–577.

Chisti, Y. (2008). Biodiesel from microalgae beats bioethanol. *Trends in Biotechnology, 26*(3), 126–131.

Chung, I. K., Oak, J. H., Lee, J. A., Shin, J. A., Kim, J. G., & Park, K. S. (2013). Installing kelp forests/seaweed beds for mitigation and adaptation against global warming: Korean project overview. *ICES Journal of Marine Science, 70*(5), 1038–1044.

Davis, T. A., Volesky, B., & Mucci, A. (2003). A review of the biochemistry of heavy metal biosorption by brown algae. *Water Research, 37*(18), 4311–4330.

Field, C. B., Barros, V., Stocker, T. F., & Dahe, Q. (2012). *Managing the Risks of Extreme Events and Disasters to Advance Climate Change Adaptation: Special Report of the Intergovernmental Panel on Climate Change*. Cambridge University Press.

Jacquin, A. G., Brulé-Josso, S., Cornish, M. L., Critchley, A. T., & Gardet, P. (2014). Selected comments on the role of algae in sustainability. In *Advances in Botanical Research* (Vol. 71, pp. 1–30). Academic Press.

Janowiak, M., Swanston, C., & Ontl, T. (2017). *Management of Forest Carbon Stocks*. U.S. Department of Agriculture, Forest Service, Climate Change Resource.

Melillo, J. M., Richmond, T. T., & Yohe, G. (2014). *Climate Change Impacts in the United States: Third National Climate Assessment* (p. 52).

Notoya, M. (2010). Production of biofuel by macroalgae with preservation of marine resources and environment. In *Seaweeds and Their Role in Globally Changing Environments* (pp. 217–228). Springer, Dordrecht.

Olguín, E. J. (2003). Phytoremediation: Key issues for cost-effective nutrient removal processes. *Biotechnology Advances, 22*(1), 81–91.

Oncel, S. S. (2013). Microalgae for a macro energy world. *Renewable and Sustainable Energy Reviews, 26*, 241–264.

Oswald, W. J., Gotaas, H. B., Golueke, C. G., Kellen, W. R., Gloyna, E. F., & Hermann, E. R. (1957). Algae in waste treatment. *Sewage and Industrial Wastes, 29*(4), 437–457.

Park, J. B. K., Craggs, R. J., & Shilton, A. N. (2011). Wastewater treatment high-rate algal ponds for biofuel production. *Bioresource Technology, 102*(1), 35–42.

Rhodes, C. J. (2010). Biofuel from algae: Salvation from peak oil? In *Seaweeds and Their Role in Globally Changing Environments* (pp. 229–248). Springer, Dordrecht.

Sing, S. F., Isdepsky, A., Borowitzka, M. A., & Moheimani, N. R. (2013). Production of biofuels from microalgae. *Mitigation and Adaptation Strategies for Global Change, 18*(1), 47–72.

Singh, J., & Dhar, D. W. (2019). Overview of carbon capture technology: Microalgal biorefinery concept and state-of-the-art. *Frontiers in Marine Science, 6*, 29.

Stephenson, P. G., Moore, C. M., Terry, M. J., Zubkov, M. V., & Bibby, T. S. (2011). Improving photosynthesis for algal biofuels: Toward a green revolution. *Trends in Biotechnology, 29*(12), 615–623.

Sutherland, D. L., Howard-Williams, C., Turnbull, M. H., Broady, P. A., & Craggs, R. J. (2013). Seasonal variation in light utilization, biomass production and nutrient removal by wastewater microalgae in a full-scale high-rate algal pond. *Journal of Applied Phycology*, 1–13.

Vose, J. M., Peterson, D. L., & Patel-Weynand, T. (2012). *Effects of Climatic Variability and Change on Forest Ecosystems: A Comprehensive Science Synthesis for the U.S. Forest Sector* (General Technical Report PNW-GTR-870). USDA Forest Service, Pacific Northwest Research Station.

CHAPTER 2

Role of Microalgae in Climate Stabilization

DHAVAL KUMAR SHRIVASTAVA,[1] ASHUTOSH SRIVASTAVA,[2,*] and NAMITA BEDI[1,*]

[1]Amity Institute of Biotechnology, Amity University Uttar Pradesh, Noida, Uttar Pradesh, India

[2]Department of Zoology, Siddharth University, Kapilvastu, Siddharth Nagar, Uttar Pradesh, India

*Corresponding authors

ABSTRACT

The concentration of greenhouse gases, that is, carbon dioxide (CO_2), methane, carbon monoxide, and so forth, is increasing day by day due to anthropogenic activities, which are affecting climatic conditions. A solution to this problem is climate stabilization, which is defined as limiting emissions and maintaining CO_2 and other greenhouse gas concentrations. Microalgae cultivation can be proposed as one of the main strategies to minimize the acceleration of the Earth's temperature and to support climate stabilization, bioenergy production, and food security. As marine algae are strongly autotrophic, their cultivation presents many challenges. This chapter discusses the advantages, challenges, and future aspects of microalgae in climate stabilization.

2.1 INTRODUCTION

With the advent of the industrial revolution, concentrations of greenhouse gases have risen substantially and are most likely responsible for 50% of

Environmental Phycology: Climate Change Mitigation and Hazardous Pollutant Cleanup Using Algal Resources. Jeyabalan Sangeetha and Devarajan Thangadurai (Eds.)

DOI: 10.1201/9781779647085-2

global warming. CO_2 is the greenhouse gas produced in the largest quantities, and its atmospheric concentration has risen about 35% since 1750. As carbon dioxide (CO_2) in the atmosphere is long-lived, it can produce severe effects on the Earth and future generations, which could result in the destabilization of the climate (National Research Council, 2011).

Climate stabilization is the process of limiting CO_2 emissions and maintaining concentrations of other greenhouse gases such as methane and nitrous oxide. The World Meteorological Organization Greenhouse Gas Bulletin reported that the globally averaged concentration of CO_2 reached 407.8 ppm in the atmosphere in 2018 (World Meteorological Organization, 2019) (Figure 2.1). It has been reported that CO_2 remains in the atmosphere for centuries and in the oceans for even longer. The National Oceanic and Atmospheric Administration (NOAA) Annual Greenhouse Gas Index shows that from 1990 to 2018, radiative forcing by long-lived greenhouse gases (LLGHGs) increased significantly, with CO_2 accounting for about 80% of this increase (Index, 2022). This means that climate change caused by CO_2 is expected to persist for centuries even if emissions were halted at any point in time. Most greenhouse gases, primarily CO_2, contribute to global warming and the subsequent rise in sea level, leading to dangerous climate disruptions such as droughts and floods. "Climate stabilization" is primarily concerned with stabilizing atmospheric concentrations of CO_2.

For climate stabilization, temperature change is a critical factor. To achieve this objective, in 1988, the Intergovernmental Panel on Climate Change (IPCC) made significant progress in identifying the actions required to prevent dangerous anthropogenic interference with the climate system. In Paris, 195 countries agreed at the 21st Conference of the Parties (COP21) to limit the increase in mean global temperature to less than 2°C above preindustrial levels (Greene et al., 2017). As CO_2 flows into and out of the oceans and biosphere as part of the planet's natural cycles, atmospheric CO_2 concentrations are expected to persist for many centuries, even if emissions were halted at any point in time. The higher the cumulative CO_2 and other greenhouse gas emissions, the greater the resulting atmospheric concentrations of these gases, thereby prolonging the duration of global warming. Prolonged warming could result in a global mean temperature rise of more than 2°C and a sea level rise of 1 to 4 meters. Climate stabilization requires stabilizing atmospheric CO_2 concentrations and regulating CO_2 emissions through natural carbon sinks.

Regulating CO_2 concentrations is essential for stabilizing the global mean temperature.

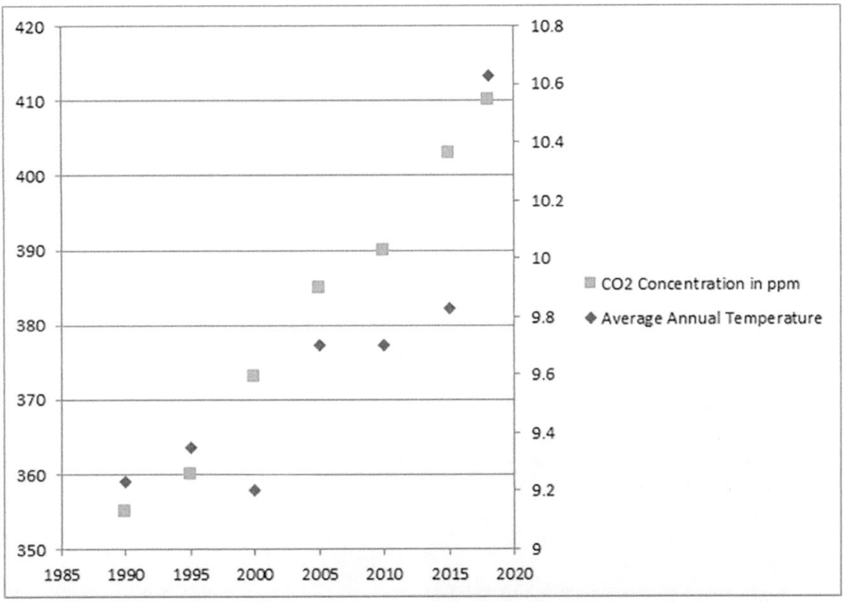

FIGURE 2.1 Average annual temperature and CO_2 concentration (ppm), 1990–2018, according to the World Meteorological Organization (2019).

2.2 SOURCES AND EFFECTS OF GREENHOUSE GASES

In 2018, the 2.0% increase in global greenhouse gas emissions was mainly due to a 2.0% increase in global fossil CO_2 emissions from fossil-fuel combustion and those from industrial non-combustion processes, including cement production. Global emissions of methane (CH_4) and nitrous oxide (N_2O) increased by 1.8% and 0.8%, respectively. Global emissions of fluorinated gases (so-called F-gases) continued to grow by an estimated 6% in 2018, thereby also contributing to the 2.0% growth in total greenhouse gas emissions. The 2.0% growth in greenhouse gas emissions is higher than the annual average increase of 1.2% since 2012 but lower than the 2.5% increase over the first decade of this century. Fossil CO_2 emissions are the largest source of global greenhouse gas emissions, with a share of about 76%, followed by CH_4 (16%), N_2O (6%), and F-gases (3%) (Figure 2.2).

The five largest emitters of greenhouse gases, together accounting for 62% of global emissions, are China (26%), the United States (13%), the

European Union (more than 8%), India (7%), the Russian Federation (5%), and Japan (almost 3%). In 2018, a significant increase in greenhouse gas emissions was observed in four of these regions: China (+1.9%), India (+5.5%), the United States (+2.7%), and the Russian Federation (+5.1%), whereas emissions decreased in the European Union (−1.5%) and Japan (−1.2%) (Lamb et al., 2022).

Since the largest contributor to greenhouse gases is CO_2, its reduction can be achieved by three different methods (Molazadeh et al., 2019): chemical reactions (approach to capturing CO_2, e.g., mineral carbonization); storage approach (the emitted CO_2 is trapped or injected into the oceans); and biological approaches for removing CO_2.

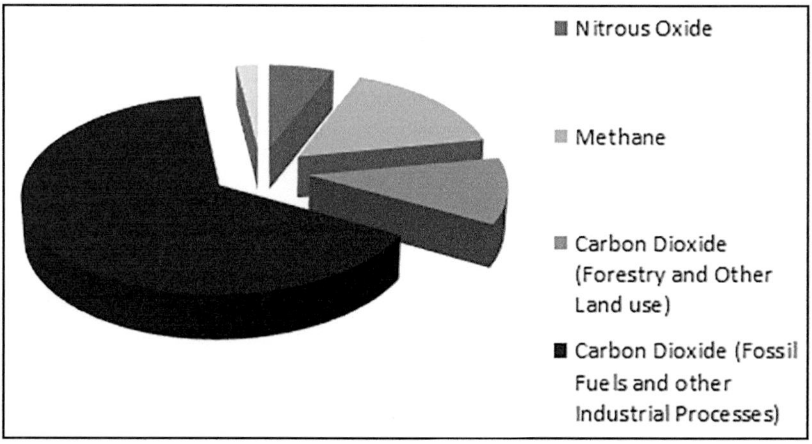

FIGURE 2.2 Global greenhouse gas emissions by gas.

2.2.1 *BIOLOGICAL APPROACHES*

1. **Forestation:** This includes the concept of afforestation, reforestation, farming of crops, and rearing of livestock. In this method, no chemicals are used, though it requires a large area and a long time and can also compete with food crops for arable land (Cheah et al., 2016).

2. **Oceanic Fertilization:** This includes the concept of fertilizing the oceans with organic and inorganic nutrients so that the CO_2 uptake efficiency of phytoplankton can be increased. Though this has significant potential for capturing CO_2 in the future, there are still some major issues such as cost optimization and unintended

impacts on marine microflora and microfauna, thus affecting marine biodiversity (Williamson et al., 2012).

3. **Microalgae-based Capture and Utilization of Carbon:** This includes the concept of bioconversion of CO_2 into biofuels and other viable products through photosynthesis. This process has many advantages such as high efficiency over a wide concentration range of CO_2, faster processes due to the faster growth rate of microalgae, no land is required, and more than one product can be obtained, that is, food, feed, and biofuel. This method has some major drawbacks as well, such as the culture system must be large, thus making the process of biomass production, harvesting of the products, as well as by-products via downstream processing expensive. There is a high chance of contamination, and the algal cells can be sensitive to NO_X and SO_X (fuel gas components), and optimum cultural conditions must always be maintained, thus making this process labor-intensive (Varshney et al., 2015).

2.3 POTENTIAL OF MICROALGAE IN THE PROCESS OF CO_2 SEQUESTRATION

Marine microalgae have been reported as a globally emerging oceanic carbon sink. They are appearing as an upcoming approach to minimize the Earth's temperature by the removal of CO_2 and simultaneous creation of inexhaustible biofuels, biomass, and other value-added by-products, for example, beauty care products, nutraceuticals, pharmaceuticals, biocompost, and biologically active compounds (Qari et al., 2017). Microalgae have the potential to decrease the liberation of CO_2 into the air by binding carbon, and this can be done through a CO_2 concentrating mechanism (CCM) for proficiently photosynthesizing by procuring inorganic carbon even from low barometric CO_2 concentrations (Le Manach et al., 2019).

Blue carbon strategies for climate stabilization include using microalgal biomass as biofuel instead of fossil fuels, as coal and other fossil fuel combustion are responsible for CO_2 emissions (Kandasamy et al., 2022). As marine microalgae are emphatically autotrophic, their capability of producing more organic matter compared to the process of ecological respiration makes them an effective CO_2 capture machinery in aquatic habitats (Krause-Jensen and Duarte, 2016).

Marine microalgae include diatoms, dinoflagellates, green algae, and cyanobacteria. Microalgal growth rates are comparatively 5–10% higher than those of other terrestrial productive plants. Additionally, they show a great ability to grow in variable environments such as saline-alkaline water, land, and wastewater. They require less arable land than terrestrial crops, which avoids competition with agriculture for food production (Zullaikah et al., 2019). Microalgae's simple cell structure and rapid growth rate increase their CO_2 fixing effectiveness over their counterparts (Khan et al., 2018). Microalgae can be used to bioremediate the greenhouse gases and other pollutants produced by industry and sewage and convert them into biomass and energy (Christie-Oleza et al., 2017).

Microalgae can capture as much as 90% of CO_2 or bicarbonate in open ponds, although the capacity varies among species. Certain groups of algae have a high tolerance for elevated concentrations of CO_2 (Cheah et al., 2015). It has been reported that *Scenedesmus* was able to tolerate extremely high CO_2 concentrations better than *Chlorella*, although both algae exhibited similar growth rates when the CO_2 concentration was in the range of 10–30%. A microalgal consortium obtained from wastewater ponds demonstrated high CO_2 sequestration (53–100%; 150–291 mg/g) (Sarwer et al., 2022).

Approximately 16.6 million tons of carbon are annually buried in aquaculture ponds around the world, with maximum sequestration occurring in Asia, particularly in China. The carbon fixing rates of microalgae are much higher than those of terrestrial plants. It is known that 1.6 to 2 grams of captured CO_2 are used for every gram of algal biomass (Jena and Das, 2011).

It was determined that the maximum cutoff for the CO_2 capture capacity of ocean macroalgal aquaculture is 2.48 million tons of CO_2 (0.68 Tg C) per year. However, this maximum threshold represents only about 0.4% of the global wild algal system's CO_2 capture capacity, estimated at 173 Tg C $year^{-1}$ (Krause-Jensen and Duarte, 2016). It has been predicted that, at these growth rates, the maximum limit of CO_2 capture capacity in oceanic aquaculture would surpass 6% of global CO_2 sequestration by 2050.

2.4 MICROALGAE SELECTION AND CHARACTERIZATION CRITERIA

A higher rate of growth and high conversion efficiency of CO_2 are the two major factors that must be kept in mind while selecting the species.

The advantage of microalgae is their uncomplicated cell structure, short life cycle, and high nutrient absorption rate, as well as their capability of growing in both fresh and marine water, including the fact that there are several species of microalgae that are possible in a miniaturized system that further reduces the cost of the system and intensifies the design of both the process and harvest. Along these lines, their high efficiency of photosynthesis and the possibility of their growth and development in either mixotrophic or heterotrophic modes are a huge advantage in comparison to other photosynthetic organisms (Morales-Sánchez et al., 2017). It also includes the selection of the best cultivation mode and identification of low-cost growth medium to increase the production of biomass, selecting a harvest method that not only maximizes nutrient recovery and reuse but also helps in the recovery of the product on a global level (Coêlho et al., 2019). Table 2.1 indicates reasons for choosing microalgae as a desired source for the production of biomass (Dębowski et al., 2020; Yanlin et al., 2014).

TABLE 2.1 Reasons for Choosing Microalgae as a Desired Source for Biomass Production

Feature	Advantages
High growth rate	• Better than other species designed for the same purpose • Requires less space
High product content	• Value of biomass yield is high
Growth in extreme environments	• Rate of contamination and rate of degradation are low
Large cell size; can be colonial or filamentous	• Cost-effective harvesting and processing
Wide tolerance to environmental conditions	• Culture conditions can be dynamic, thus allowing the growth of microalgae in a vast range of climatic conditions
High CO_2 tolerance and absorption	• Optimized use of CO_2 residues, thus improving potential for CO_2 sequestration
Shear force tolerance	• Pumping methods are inexpensive
Tolerance to contaminants	• Can be grown in polluted water and in the presence of greenhouse gases
Does not produce self-inhibitors	• When biomass density is high, there is minimal inhibition of growth

Source: From Griffiths et al. (2009) and Klein et al. (2018).

2.5 CO_2 SEQUESTRATION BY MICROALGAE: SYSTEM AND PATHWAY

In adapting to conditions of low ambient CO_2 concentration and increasing demand for inorganic carbon, numerous microalgae have developed mechanisms that utilize energy to elevate CO_2 concentrations in the vicinity of RuBisCo. These components are known as CO_2-concentrating mechanisms (CCMs). The steady-state concentration of CO_2 in carboxysomes is increased by the CO_2 generated by carbonic anhydrase (CA), compared with that in the surrounding medium, thereby promoting carboxylase activity rather than the oxygenase activity of RuBisCo.

Hence, CCMs generally include (i) active transporters for bicarbonate uptake; (ii) a set of strategically localized carbonic anhydrases (CAs); and (iii) confinement of RuBisCo within a subcellular microcompartment (Mistry et al., 2019). Figure 2.3 presents a schematic representation of the CO_2 sequestration mechanism in alpha- and beta-cyanobacterial cells, and Figure 2.4 shows the Calvin–Benson cycle.

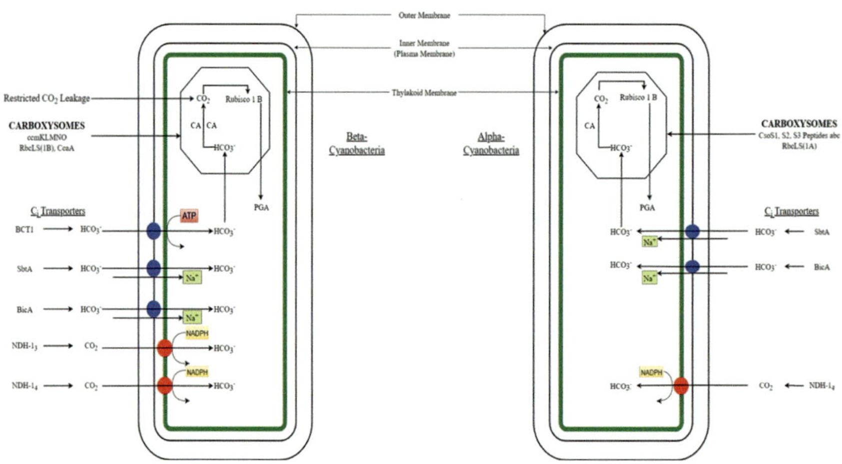

FIGURE 2.3 Schematic representation of the CO_2 sequestration mechanism in alpha- and beta-cyanobacterial cells.

2.6 MICROALGAE AS A POTENTIAL ALTERNATIVE FOR WASTEWATER TREATMENT AND CLIMATE STABILIZATION

Each year, almost 1000 km^3 of wastewater is produced worldwide, which incorporates more than 300 km^3 of municipal wastewater and more than

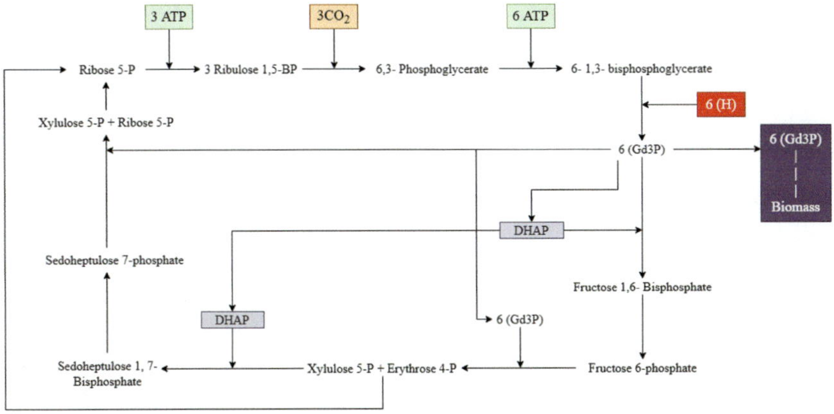

FIGURE 2.4 Calvin–Benson cycle.

600 km³ from industries. While developed nations treat 70% of their wastewater, only 8% of the same is treated by developing and underdeveloped nations (Kumar et al., 2019). WWTPs have been recognized as one of the largest producers of greenhouse gases causing global warming. There are two methods of GHG production: one is direct GHG released by the management of sludge and wastewater collection systems, and the other is the indirect method from the disposal of waste and importation of electrical and thermal energy (Bani Shahabadi et al., 2009).

According to an assessment, approximately 0.77 Gt CO_2 greenhouse gas (GHG) emissions were produced due to the deterioration of organic matter during wastewater treatment (WWT), which was proportional to about 1.57% of global GHG emissions, that is, 49 Gt CO_2 emissions (IPCC, 2014). WWT accounts for 0.56–0.71 Gt CO_2 emissions annually between 2005 and 2030 and non-CO_2 GHG emissions representing approximately 4.6–5.2% of global non-CO_2 GHG emissions (USEPA, 2012). Although the direct release of CO_2 from the deterioration of wastewater organics can be considered a carbon-neutral procedure, the concern is the increase in non-CO_2 GHG emissions, which are 25–298 times more potent in their greenhouse effect than CO_2 (USEPA, 2020).

Microalgae are emerging as a strategy for utilizing wastewater for active and direct CO_2 capture through carbon capture and utilization (CCU). Since CCU can occur within preexisting wastewater systems during the treatment process, no additional land or logistics are required for these activities. Consequently, integrating CCU with WWT may transform energy-intensive, carbon-emitting wastewater treatment plants (WWTPs) into integrated water resource recovery facilities that recover energy, nutrients, water, and other valuable carbon-based products with economic, environmental, and societal benefits.

2.7 MICROALGAE-ENABLED CCU-BASED WASTEWATER TREATMENT METHODS

The customary goal of WWT is to remove various contaminants (e.g., organic waste, inorganic waste, and pathogens) so that the treated effluent meets ecological quality standards. In this context, microalgae-enabled CCU must be implemented without compromising treatment efficacy, given that the protection of the surrounding environment and the safety of public health are of utmost importance for the wastewater industry (Salek et al., 2013).

1. **High-Rate Algal Ponds (HRAP):** This is a combined approach for WWT and algal biomass production. It is a mutualistic system in which microalgae growing in oxidation ponds provide oxygen for bacteria, and in turn, bacteria supply nutrients for algal growth. The system is highly cost-efficient and reduces biological oxygen demand (BOD). Significant potential has been demonstrated by microalgal genera such as *Oscillatoria*, *Micractinium*, *Arthrospira*, and *Scenedesmus* to grow in HRAP systems and are easier to harvest compared with smaller single-celled algal species such as *Chlorella* sp. and *Oocystis* sp. (Molazadeh et al., 2019).
2. **Cultivation of Microalgae:** The actual productivity of microalgae is higher than that of terrestrial plants, with reproductive rates generally in the range of 15–30 g m^{-2} d^{-1} (Moody et al., 2014). Microalgae fix 1.8–2.4 kg of CO_2 per kg of biomass produced and yield approximately 9.4–116 g of CO_2 fixed per gram of nitrogen absorbed (C/N ratio = 2.6–32) when cultivated autotrophically (Li et al., 2017) (Figure 2.5). These high-efficiency microalgal systems utilize suspended micronutrients, that is, nitrogen and phosphorus from wastewater, as well as improved strains and cultures that achieve reliable treatment performance by creating specific conditions through process engineering (Hu et al., 2017). Overall, the process has major advantages, such as high CCU efficiency with simultaneous nutrient removal, production of biomass for generating

biofuels and various bioproducts, polishing of organics, and feasibility for full-scale application. At the same time, the process also has major limitations, including unstable performance, high land requirements, high costs, high energy demand during harvesting, and limited removal of organics (Lu et al., 2018).

3. **Generation of Biochar by Microalgae:** Algal biochar has a high nutrient content, and when used as a soil amendment, it improves soil fertility and water retention capacity, which can lead to higher crop yields (Bird et al., 2012). Algal biochar can also be used as a tool for water remediation and stable soil carbon sequestration. It integrates photosynthesis with an enhanced carbon cycle to increase soil carbon storage and reduce soil methane and nitrogen emissions. The production of biochar from *Spirulina* sp. occurs via the slow pyrolysis technique at temperatures of 450 to 600°C. The maximum yield of biochar occurs at approximately 500 and 550°C (Chaiwong et al., 2013).

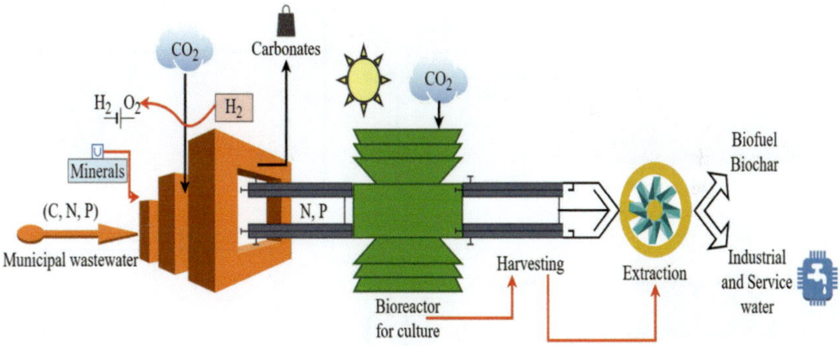

FIGURE 2.5 Microalgae-enabled CCU-based wastewater treatment system.

2.7.1 CONCEPT OF CO-LOCATION

Integrating CCU with the WWT process not only changes the operational framework and structure of water resource recovery facilities (WRRFs) but also creates opportunities for major CO_2-producing industries, such as power plants, cement plants, and processing facilities. CO_2 from the source can be captured by WRRFs to produce concentrated carbon products and value-added by-products, which are likely to be more cost-effective than independent CCU implementation by power plants (USEPA, 2020). Ca/Mg-rich fly ash generated by power plants can be used as a silicate supplement to promote the mineralization of CO_2 (Song et al., 2016).

The WRRF-treated wastewater can be reused by power plants, and WRRF-generated renewable energy can be integrated into the power grid. The products generated—including biofuels, biochar, $CaCO_3$, and filtered fly ash—have the potential to benefit both industries. Furthermore, the feasibility of CCU has been demonstrated using both MECC and microalgae. Economically and industrially available ion-exchange resins operating through a dry–wet cycle can concentrate ambient CO_2 by more than 200 times for MECC to produce concentrated H_2 and carbonate, and similar practices can be used to enhance microalgal yield and biofuel production. Moreover, CCU-enabled WWT practices, such as municipal WRRFs, can be implemented in numerous industries where WWT is required. A schematic representation of the mutual benefits shared between a coal-fired power plant (a point source) and a MECC–microalgae-enabled CCU–based WWT system is shown in Figure 2.6.

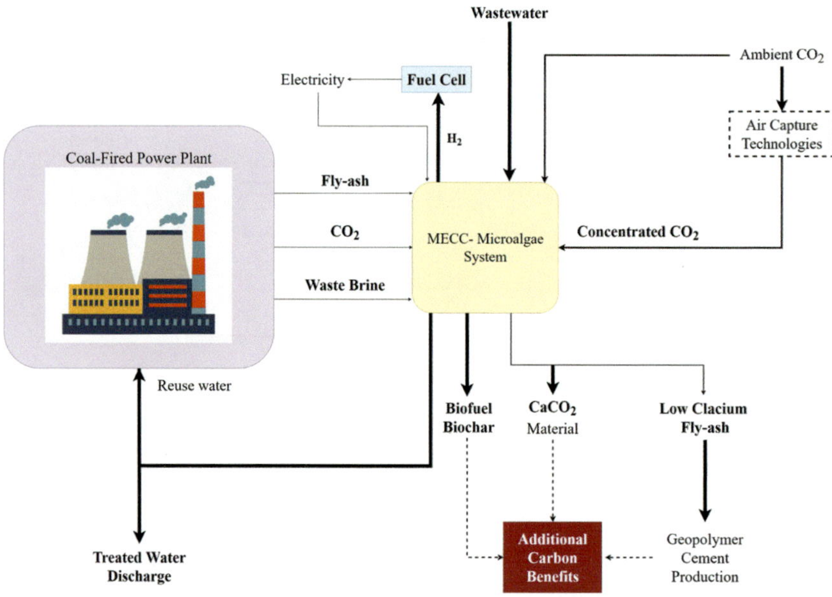

FIGURE 2.6 Schematic representation of the mutual benefits shared between a coal-fired power plant (i.e., a point source) and a MECC–microalgae-enabled CCU-based wastewater treatment system.

2.8 MICROALGAE AS PROMISING BIOFUEL AND BIOMASS AS FOOD

Algal biomass produces value-added products in bulk form and enhances the total economic efficiency of algae (Moncada et al., 2015). Microalgae are designated candidates in biorefinery structures as they have the capability to generate a variety of important items (Mutanda et al., 2020). When fast development and challenges in the flexibility of the food chains are considered, they are an important asset of biomass.

Extraction of fuel-based products from the biomass is done by standard heat destruction methods in thermochemical conversion. Examples of the same include gasification, which generates syngas; thermal liquefaction, which generates bio-oil; pyrolysis, which generates syngas, bio-oil, and biochar; and direct combustion/burning, which produces heat and electricity. The bioconversion of energy into biofuels via natural bioprocessing is represented through the process of biochemical transformation. Examples of biochemical conversion include anaerobic assimilation and digestion, which generate biogas; alcoholic maturation/fermentation, which generates bioethanol; and the production of bio-hydrogen. The transesterification reaction, which includes the reaction between algal triglycerides and alcohol in the presence of a catalyst, also leads to the generation of biodiesel.

Another significant angle of this concept is the microbial energy components, that is, microbial fuel cells (MFCs), which are essentially bio-electrochemical contraptions that can generate power/energy or bioelectricity from the electrochemical biodegradation of organic complexes under hypoxic conditions. The amalgamation of microalgae in this microbial energy component has been viewed as having huge potential in the production of oxygen as well, because of the removal of CO_2 by photosynthesis via algal development (Shaw and Mukherjee, 2022).

Microalgae can effectively expel CO_2 through the rapid production of algal biomass, which is emerging as an efficient alternative protein source. Proteins based on microalgae can effectively meet the population's requirement for protein, with a few points of interest over others, one of which is decreasing GHGs from domesticated animals. Each year, livestock production pumps out greenhouse gases with the equivalent warming effect of more than 7 gigatons of CO_2; roughly 40% of that is produced during digestion, as cattle, goats, and sheep belch and pass methane, a greenhouse gas. Other than proteins, microalgae are a source of several important compounds with medical benefits, for example, starches, polyunsaturated fatty acids, essential minerals, and nutrients, which can enhance the nutritional value of food items after their incorporation. Polysaccharides and oligosaccharides are promising compounds with potential medical benefits, with emerging considerations regarding prebiotic applications. Lipids, specifically polyunsaturated long-chain omega-3 fatty acids, have also been recommended as important compounds with medical benefits that can be incorporated into food items. α-linolenic acid, eicosapentaenoic acid, docosapentaenoic acid, and docosahexaenoic acid are some of the most significant ω-3 PUFAs with medical benefits for humans (Caporgno and Mathys, 2018) (Figure 2.7).

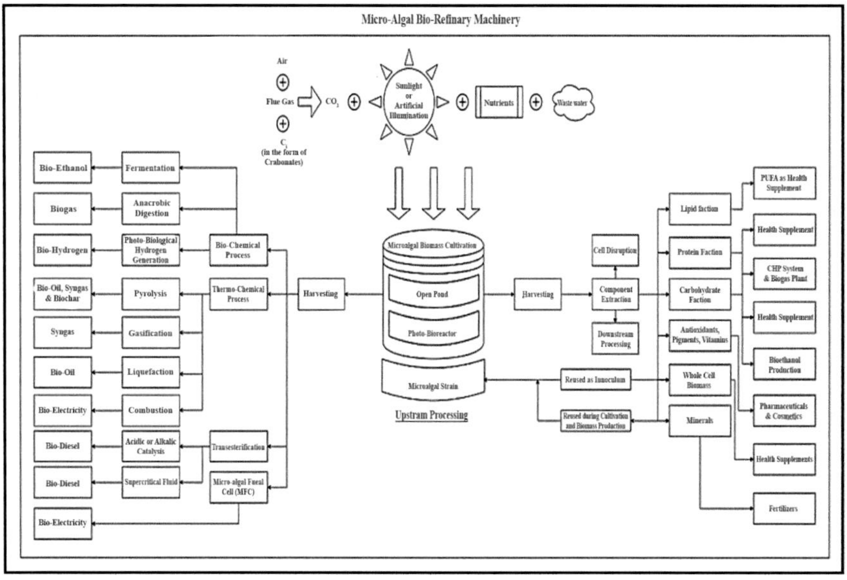

FIGURE 2.7 Summary of microalgae as a source of biofuels and other value-added products.

The generation of biofuels from marine microalgae can be a significant atmospheric adjustment and climate stabilization procedure in lessening society's reliance on petroleum products (Johnson et al., 2021). The IPCC has found that emissions from fossil fuels are the dominant cause of global warming. In 2018, 89% of global CO_2 emissions came from fossil fuels and industry. There is a need for carbon-neutral biofuels.

When such techniques are created to deliver carbon-neutral biofuels, they can be further altered to accomplish negative emissions through the generation of enduring bio-oil-based commodities. By utilizing microalgae-based bio-oil as a feedstock in the synthesis of widely utilized chemical substances, for example, plastics (Otto et al., 2015; Zeller et al., 2013), industry can accomplish negative emissions while earning revenue at the same time. As opposed to the geographical stockpiling of captured CO_2, utilization of these plastics and other bio petroleum-based items in development extending on a broad scale can give us a more secure and economically worthwhile technique for sequestering a lot of carbon for extended timeframes (Granados et al., 2016). Marine microalgae play a direct role in relieving and potentially turning around the impacts of fossil CO_2 emissions and can likewise play a similarly significant, yet circuitous, job in atmosphere alleviation. As previously referenced, the primary production rate of microalgae is higher than terrestrial plants, thus reducing the land area required to deliver an equivalent measure of bioenergy as well as food. By substituting the development of marine microalgae for conventional agricultural practices, significant CO_2 emissions and water reserves can be accomplished. Additionally, on a worldwide scale, the evaded emissions coming from the dislodging of traditional agricultural practices may surpass the advantages of micro-algal biofuels in accomplishing atmosphere adjustment objectives (Beal et al., 2015; Granados et al., 2016).

Microalgae-based fuels are assessed to be conservative in comparison to unrefined oil, which sells for more than 100 USD per barrel. Although algal biofuels are not yet cost-effective, there are numerous organizations in the USA, Europe, and different parts of the world that are generating algae-based fuels at an industrial scale. As per the TEA model, biodiesel from algal biomass can sell beneath \$5/gallon equivalent to gasoline and bioethanol at the cost of \$2.95/gallon, which makes it economically feasible. A few evaluations suggest that the attainable and practical cost of algal biofuels in comparison with other biofuels is \$1/L.

2.9 CHALLENGES, OPPORTUNITIES, AND FUTURE TRENDS IN THE CULTIVATION OF MICROALGAE FOR CLIMATE STABILIZATION

As microalgae are autotrophic microorganisms, they require solar energy, CO_2, and nutrients for their life cycle (Granados et al., 2016). The challenges associated with the cultivation of microalgae are discussed in Table 2.2. The most promising locations for the cultivation of marine microalgae include coastal areas of Australia, Brazil, India, Mexico, the Middle East, Saharan North Africa, and southern Africa (Moody et al., 2014). Overall, large-scale production of marine microalgae can assume multiple roles in serving communities to achieve climate adaptation or climate stabilization objectives agreed upon at COP21 in Paris. Governments, as well as private stakeholders, must invest in research and development of microalgae cultivation to enhance the production of bioenergy and food while ensuring cost-effectiveness, reducing land requirements, and minimizing CO_2 emissions.

TABLE 2.2 Challenges in the Growth of Microalgae

Factors	Description	Effect
Light	Duration and intensity	• Affects the growth rate, biochemical composition of microalgae, and biomass yield. • The optimum light intensity for some microalgal species is 100 μmol m^{-2} s^{-1}.
Temperature	Optimum temperature: 20–30°C	• Directly influences biochemical processes, including photosynthesis. • Any deviation from the optimum range retards or even stops the growth and activity of algae.
Macronutrients (nitrogen, phosphorus, carbon, silicon)	Wastewater treatment systems help in the efficient recycling of nutrients, especially phosphorus. This integration is a long-term solution for nutrient recycling.	• Deficiency affects the growth rate as well as biomass quantity and the concentration of carbohydrates and lipids in microalgae.
Carbon dioxide	Crosses the air–water interface	• Can limit the growth rate even at low concentrations. • The concentration of CO_2 at the cultivation site depends on fossil- and industrially derived CO_2, as well as its transportation offsite. • Direct capture of CO_2 at the cultivation site can be carried out in a cost-effective manner to overcome this issue.
Economics	Price decreasing by 1–2% per year	• Leads to reduced interest among farmers and investors. • Economic compensation should be provided by the government to enhance long-term benefits.

Source: Greene et al. (2016).

2.10 CONCLUSION

Carbon sequestration through microalgae offers a promising approach for reducing global CO_2 emissions. Although many studies have focused on advanced cultivation techniques, harvesting methods, and the CO_2 sequestration capacity of microalgae, there is still a need to scale up their application and commercialization to take advantage of favorable growth conditions and enhance carbon dioxide capture capacity. The CO_2 tolerance, carbon assimilation efficiency, photosynthetic performance, and growth rates of algal cells are higher than those of terrestrial plants, making microalgae a more suitable option for commercial, cost-effective, open-system carbon sequestration.

KEYWORDS

- **biological oxygen demand**
- **carbon capture**
- **carbon dioxide**
- **carbon sequestration**
- **carboxylase activity**
- **climate stabilization**
- **greenhouse gases**
- **microalgae**

REFERENCES

Bani Shahabadi, M., Yerushalmi, L., & Haghighat, F. (2009). Impact of process design on greenhouse gas (GHG) generation by wastewater treatment plants. *Water Research, 43*(10), 2679–2687. https://doi.org/10.1016/j.watres.2009.02.040 (accessed on 20th February 2026).

Beal, C. M., Gerber, L. N., Sills, D. L., Huntley, M. E., Machesky, S. C., Walsh, M. J., et al. (2015). Algal biofuel production for fuels and feed in a 100-ha facility: A comprehensive techno-economic analysis and life cycle assessment. *Algal Research, 10*, 266–279. https://doi.org/10.1016/j.algal.2015.04.017 (accessed on 20th February 2026).

Bird, M. I., Wurster, C. M., De Paula Silva, P. H., Paul, N. A., & De Nys, R. (2012). Algal biochar: Effects and applications. *GCB Bioenergy, 4*(1), 61–69. https://doi.org/10.1111/j.1757-1707.2011.01109.x (accessed on 20th February 2026).

Caporgno, M. P., & Mathys, A. (2018). Trends in microalgae incorporation into innovative food products with potential health benefits. *Frontiers in Nutrition, 5*, 58. https://doi.org/10.3389/fnut.2018.00058 (accessed on 20th February 2026).

Chaiwong, K., Kiatsiriroat, T., Vorayos, N., & Thararax, C. (2013). Study of bio-oil and bio-char production from algae by slow pyrolysis. *Biomass and Bioenergy, 56*, 600–606. https://doi.org/10.1016/j.biombioe.2013.05.035 (accessed on 20th February 2026).

Cheah, W. Y., Ling, T. C., Juan, J. C., Lee, D. J., Chang, J. S., & Show, P. L. (2016). Biorefineries of carbon dioxide: From carbon capture and storage (CCS) to bio energies production. *Bioresource Technology, 215*, 346–356. https://doi.org/10.1016/j.biortech.2016.04.019 (accessed on 20th February 2026).

Cheah, W. Y., Show, P. L., Chang, J. S., Ling, T. C., & Juan, J. C. (2015). Bio sequestration of atmospheric CO_2 and flue gas-containing CO_2 by microalgae. *Bioresource Technology, 184*, 190–201. https://doi.org/10.1016/j.biortech.2014.11.026 (accessed on 20th February 2026).

Christie-Oleza, J. A., Sousoni, D., Lloyd, M., Armengaud, J., & Scanlan, D. J. (2017). Nutrient recycling facilitates long-term stability of marine microbial phototroph-heterotroph interactions. *Nature Microbiology, 2*, 17100. https://doi.org/10.1038/nmicrobiol.2017.100 (accessed on 20th February 2026).

Coêlho, D. d. F., Tundisi, L. L., Cerqueira, K. S., Rodrigues, J. R. d. S., Mazzola, P. G., Tambourgi, E. B., et al. (2019). Microalgae: Cultivation aspects and bioactive compounds. *Brazilian Archives of Biology and Technology, 62*. http://old.scielo.br/scielo.php?script=sci_arttext&pid=S1516-89132019000101000&nrm=iso (accessed on 20th February 2026).

Dębowski, M., Zieliński, M., Kazimierowicz, J., Kujawska, N., & Talbierz, S. (2020). Microalgae cultivation technologies as an opportunity for bioenergetic system development—Advantages and limitations. *Sustainability, 12*(23), 9980. https://www.mdpi.com/2071-1050/12/23/9980 (accessed on 20th February 2026).

Greene, C. H., Huntley, M. E., Archibald, I., Gerber, L. N., Sills, D. L., Granados, J., et al. (2017). Geoengineering, marine microalgae, and climate stabilization in the 21st century. *Earth's Future, 5*(3), 278–284. https://doi.org/10.1002/2016EF000486 (accessed on 20th February 2026).

Greene, C. H., Huntley, M. E., Archibald, I., Gerber, L. N., Sills, D. L., Granados, J., et al. (2016). Marine microalgae: Climate, energy, and food security from the sea. *Oceanography, 29*(4), 10–15. https://doi.org/10.5670/oceanog.2016.91 (accessed on 20th February 2026).

Hu, Y., Hao, X., van Loosdrecht, M., & Chen, H. (2017). Enrichment of highly settleable microalgal consortia in mixed cultures for effluent polishing and low-cost biomass production. *Water Research, 125*, 11–22. https://doi.org/10.1016/j.watres.2017.08.034 (accessed on 20th February 2026).

Index, A. G. G. (2022). The NOAA annual greenhouse gas index (AGGI). https://gml.noaa.gov/aggi/aggi.html (accessed on 20th February 2026).

IPCC. (2014). *Climate Change 2014: Impacts, Adaptations, and Vulnerability. Fifth assessment report of the Intergovernmental Panel on Climate Change.* https://www.ipcc.ch/report/ar5/wg2/ (accessed on 20th February 2026).

Jena, U., & Das, K. C. (2011). Comparative evaluation of thermochemical liquefaction and pyrolysis for bio-oil production from microalgae. *Energy and Fuels, 25*(11), 5472–5482. https://doi.org/10.1021/ef201373m (accessed on 20th February 2026).

Johnson, N., Gross, R., & Staffell, I. (2021). Stabilization wedges: Measuring progress towards transforming the global energy and land use systems. *Environmental Research Letters, 16*(6), 064011. https://doi.org/10.1088/1748-9326/abec06 (accessed on 20th February 2026).

Kandasamy, S., Bhuvanendran, N., Narayanan, M., & He, Z. (2022). Thermochemical conversion of algal biomass. In M. El-Sheekh & A. E. F. Abomohra (Eds.), *Handbook of Algal Biofuels* (pp. 281–302). Elsevier.

Khan, M. I., Shin, J. H., & Kim, J. D. (2018). The promising future of microalgae: Current status, challenges, and optimization of a sustainable and renewable industry for biofuels, feed, and other products. *Microbial Cell Factories, 17*(1), 36. https://doi.org/10.1186/s12934-018-0879-x (accessed on 20th February 2026).

Krause-Jensen, D., & Duarte, C. M. (2016). Substantial role of macroalgae in marine carbon sequestration. *Nature Geoscience, 9*, 737–742.

Kumar, P. K., Krishna, S. V., Naidu, S. S., Verma, K., Bhagawan, D., & Himabindu, V. (2019). Biomass production from microalgae *Chlorella* grown in sewage, kitchen wastewater using industrial CO_2 emissions: Comparative study. *Carbon Resources Conversion, 2*(2), 126–133. https://doi.org/10.1016/j.crcon.2019.06.002 (accessed on 20th February 2026).

Lamb, W. F., Grubb, M., Diluiso, F., & Minx, J. C. (2022). Countries with sustained greenhouse gas emissions reductions: An analysis of trends and progress by sector. *Climate Policy, 22*(1), 1–17. https://doi.org/10.1080/14693062.2021.1990831 (accessed on 20th February 2026).

Le Manach, S., Duval, C., Marie, A., Djediat, C., Catherine, A., Edery, M., et al. (2019). Global metabolomic characterizations of *Microcystis spp.* highlights clonal diversity in natural bloom-forming populations and expands metabolite structural diversity. *Frontiers in Microbiology, 10*, 791. https://doi.org/10.3389/fmicb.2019.00791 (accessed on 20th February 2026).

Li, Y., Leow, S., Fedders, A. C., Sharma, B. K., Guest, J. S., & Strathmann, T. J. (2017). Quantitative multiphase model for hydrothermal liquefaction of algal biomass. *Green Chemistry, 19*(4), 1163–1174. https://doi.org/10.1039/C6GC03294J (accessed on 20th February 2026).

Lu, L., Guest, J. S., Peters, C. A., Zhu, X., Rau, G. H., & Ren, Z. J. (2018). Wastewater treatment for carbon capture and utilization. *Nature Sustainability, 1*(12), 750–758. https://doi.org/10.1038/s41893-018-0187-9 (accessed on 20th February 2026).

Mistry, A. N., Ganta, U., Chakrabarty, J., & Dutta, S. (2019). A review on biological systems for CO_2 sequestration: Organisms and their pathways. *Environmental Progress and Sustainable Energy, 38*(1), 127–136. https://doi.org/10.1002/ep.12946 (accessed on 20th February 2026).

Molazadeh, M., Ahmadzadeh, H., Pourianfar, H. R., Lyon, S., & Rampelotto, P. H. (2019). The use of microalgae for coupling wastewater treatment with CO_2 bio fixation. *Frontiers in Bioengineering and Biotechnology, 7*, 42. https://doi.org/10.3389/fbioe.2019.00042 (accessed on 20th February 2026).

Moncada, J., Cardona, C. A., & Rincón, L. E. (2015). Design and analysis of a second and third generation biorefinery: The case of castor bean and microalgae. *Bioresource Technology, 198*, 836–843. https://doi.org/10.1016/j.biortech.2015.09.077 (accessed on 20th February 2026).

Moody, J. W., McGinty, C. M., & Quinn, J. C. (2014). Global evaluation of biofuel potential from microalgae. *Proceedings of the National Academy of Sciences of the United States of America, 111*(23), 8691–8696. https://doi.org/10.1073/pnas.1321652111 (accessed on 20th February 2026).

Morales-Sánchez, D., Martinez-Rodriguez, O. A., & Martinez, A. (2017). Heterotrophic cultivation of microalgae: Production of metabolites of commercial interest. *Journal of Chemical Technology and Biotechnology, 92*(5), 925–936. https://doi.org/10.1002/jctb.5115 (accessed on 20th February 2026).

Mutanda, T., Naidoo, D., Bwapwa, J. K., & Anandraj, A. (2020). Biotechnological applications of microalgal oleaginous compounds: Current trends on microalgal bioprocessing of products. *Frontiers in Energy Research, 8*. https://doi.org/10.3389/fenrg.2020.598803 (accessed on 20th February 2026).

National Research Council. (2011). *Climate Stabilization Targets: Emissions, Concentrations, and Impacts Over Decades to Millennia*. https://www.nationalacademies.org/read/12877/chapter/4#13 (accessed on 20th February 2026).

Otto, B., Beuchel, C., Liers, C., Reisser, W., Harms, H., & Schlosser, D. (2015). Laccase-like enzyme activities from Chlorophyceae green algae with potential for bioconversion of phenolic pollutants. *FEMS Microbiology Letters, 362*(11). https://doi.org/10.1093/femsle/fnv072 (accessed on 20th February 2026).

Qari, H., Rehan, M., & Nizami, A. S. (2017). Key issues in microalgae biofuels: A short review. *Energy Procedia, 142*, 898–903. https://doi.org/10.1016/j.egypro.2017.12.144 (accessed on 20th February 2026).

Salek, S. S., Kleerebezem, R., Jonkers, H. M., Witkamp, G. J., & van Loosdrecht, M. C. (2013). Mineral CO_2 sequestration by environmental biotechnological processes. *Trends in Biotechnology, 31*(3), 139–146. https://doi.org/10.1016/j.tibtech.2013.01.005 (accessed on 20th February 2026).

Sarwer, A., Hamed, S. M., Osman, A. I., Jamil, F., Al-Muhtaseb, A. A. H., Alhajeri, N. S., et al. (2022). Algal biomass valorization for biofuel production and carbon sequestration: A review. *Environmental Chemistry Letters, 20, 2797–2851*. https://doi.org/10.1007/s10311-022-01458-1 (accessed on 20th February 2026).

Shaw, R., & Mukherjee, S. (2022). The development of carbon capture and storage (CCS) in India: A critical review. *Carbon Capture Science and Technology, 2*, 100036. https://doi.org/10.1016/j.ccst.2022.100036 (accessed on 20th February 2026).

Song, X., Pan, G., Zhang, C., Zhang, L., & Wang, H. (2016). Effects of biochar application on fluxes of three biogenic greenhouse gases: A meta-analysis. *Ecosystem Health and Sustainability, 2*(2), e01202. https://doi.org/10.1002/ehs2.1202 (accessed on 20th February 2026).

USEPA. (2012). *Global Anthropogenic Non-CO_2 Greenhouse Gas Emissions: 1990–2030*. https://www.epa.gov/global-mitigation-non-co2-greenhouse-gases/global-non-co2-ghg-emissions-1990-2030 (accessed on 20th February 2026).

USEPA. (2020). *Inventory of U.S. Greenhouse Gas Emissions and Sinks: 1990–2015*. https://www.epa.gov/ghgemissions/inventory-us-greenhouse-gas-emissions-and-sinks-1990-2015 (accessed on 20th February 2026).

Varshney, P., Mikulic, P., Vonshak, A., Beardall, J., & Wangikar, P. P. (2015). Extremophilic micro-algae and their potential contribution in biotechnology. *Bioresource Technology,*

184, 363–372. https://doi.org/10.1016/j.biortech.2014.11.040 (accessed on 20th February 2026).

Williamson, P., Wallace, D. W. R., Law, C. S., Boyd, P. W., Collos, Y., Croot, P., et al. (2012). Ocean fertilization for geoengineering: A review of effectiveness, environmental impacts and emerging governance. *Process Safety and Environmental Protection, 90*(6), 475–488. https://doi.org/10.1016/j.psep.2012.10.007 (accessed on 20th February 2026).

World Meteorological Organization. (2019). *Greenhouse Gas Concentrations in Atmosphere Reach Yet Another High.* https://public.wmo.int/en/media/press-release/greenhouse-gas-concentrations-atmosphere-reach-yet-another-high (accessed on 20th February 2026).

Yanlin, Z., Cheng, Y., Junhan, L., & Guangrong, H. (2014). Lipid production by a CO_2-tolerant green microalga, *Chlorella sp.* MRA-1. *Journal of Microbiology and Biotechnology, 24*(5), 683–689. https://doi.org/10.4014/jmb.1308.08050 (accessed on 20th February 2026).

Zeller, M. A., Hunt, R., Jones, A., & Sharma, S. (2013). Bioplastics and their thermoplastic blends from *Spirulina* and *Chlorella* microalgae. *Journal of Applied Polymer Science, 130*(5), 3263–3275. https://doi.org/10.1002/app.39559 (accessed on 20th February 2026).

Zullaikah, S., Utomo, A. T., Yasmin, M., Ong, L. K., & Ju, Y. H. (2019). Eco fuel conversion technology of inedible lipid feedstocks to renewable fuel. In K. Azad (Ed.), *Advances in Eco-Fuels for a Sustainable Environment* (pp. 237–276). Woodhead Publishing.

CHAPTER 3

Utilization of Algae as Saviors to Minimize Carbon Footprint and Combat Global Climate Change

DHARITRI BORGOHAIN,[1,*] PROPALIKA BORPATRA GOHAIN,[1] and BHABEN TANTI[2]

[1]*Department of Botany, North Lakhimpur University, Lakhimpur, Assam, India*

[2]*Department of Botany, Gauhati University, Guwahati, Assam, India*

*Corresponding author

ABSTRACT

One of the foremost global threats is climate change and its far-reaching consequences. Carbon dioxide (CO_2), the most commonly emitted greenhouse gas, is a primary driver of global warming. The Earth's oceans can absorb enormous amounts of carbon through algae, vegetation, and coral. Photosynthetic eukaryotic algae act as carbon sinks. They are considered highly significant as a viable alternative for many costly climate change solutions and can help achieve the global goal of net-zero carbon emissions, as photosynthesis is one of the most sustainable strategies for capturing and storing CO_2 from the atmosphere. Microalgae have been used to capture CO_2, which can then be converted into sustainable alternatives such as biogas, bio-oil, biochar, and syngas. Through their rapid growth, algae can cover more surface area and require less water; however, they consume more CO_2. Given their relatively small size, algae can be more easily controlled in bioreactors, where they are dehydrated and ultimately utilized as fuel or biomass. Although the development of cost-competitive

Environmental Phycology: Climate Change Mitigation and Hazardous Pollutant Cleanup Using Algal Resources. Jeyabalan Sangeetha and Devarajan Thangadurai (Eds.)

DOI: 10.1201/9781779647085-3

algal biofuel production will require long-term research and development, steady progress can be made to integrate algal biofuels as an alternative to fossil fuels in everyday life, thereby contributing to the reduction of negative environmental impacts.

3.1 INTRODUCTION

3.1.1 CLIMATE CHANGE AND GLOBAL WARMING

Climate change is a natural phenomenon; however, human activities, particularly the use of fossil fuels for energy, have drastically altered this process. Climate change is defined as any long-term alteration in climate, whether resulting from natural variability or human activity. Anthropogenic emissions of greenhouse gases (GHGs), chiefly CO_2, CH_4, N_2O, and O_3, are already affecting natural and human systems worldwide, leading to warming of the Earth's surface. Atmospheric concentrations of anthropogenic CO_2, CH_4, and N_2O have increased by 30%, 145%, and 15%, respectively, over the past 250 years. These emissions are primarily due to fossil fuel use, although land use change and agriculture are also significant sources.

Climate change is a global fact. It is a qualitatively different and very critical addition to the continuum of humanity's rising environmental risks. It is only one of the wider series of destabilizing large-scale environmental changes currently occurring, each of which represents the growing human control of the environment (McMichael, 2001). Stratospheric ozone depletion, climate change, groundwater depletion, loss of biodiversity, global soil destruction, disruption of the elemental nitrogen and sulfur cycles, and proliferation of persistent organic contaminants are numerous significant global environmental shifts. These have potential effects on the sustainability of environmental systems, the welfare of the human population, the development of food, and the economic practices of human beings (Patz et al., 2000). The rising emission of greenhouse gases by humans is expected to cause a long-term change in the world's environment, and the accumulation of these gases in the lower atmosphere has widely contributed to the recent upward trend in global average temperature. The atmosphere, ocean, land, ice, and biosphere together constitute the system of the environment (Houghton, 2009). Climate change involves an exceedingly complex array of dynamic processes, with specific combinations or interactions. Climate change in various parts of the globe shows that it is likely to vary a great deal from place to place. Global climate change also involves stratospheric ozone depletion, which leads to increased atmospheric CO_2 levels and biologically damaging UV radiation reaching the Earth's surface. Growing CO_2 concentrations contribute to sea-level rise and alteration of ocean chemistry (Hallock, 2005). Environmental global changes are occurring, causing changes in both terrestrial and marine ecosystems (Walther et al., 2002). Global warming during the 20th century led to an increase in the average surface temperature of the Earth, exceeding the range of natural variability observed over previous centuries. The Intergovernmental Panel on Climate Change (IPCC) released its Third Assessment Report, which projected that the global average surface temperature would increase by 1.4 to 5.8°C over the 21st century and concluded that warming of the Earth contributes to regional climate change, thereby influencing many physical, chemical, and biological processes (Treydte et al., 2006; Walther et al., 2002).

In varying intensities, global warming will affect every human being on the globe because human actions change the chemical composition of the atmosphere by producing greenhouse gases. Researchers have calculated that with global temperatures rising, there will be changes in precipitation patterns, melting of ice caps, and a rise in sea levels, which will affect water supplies and damage the world's forests, spreading disease and causing both floods and droughts, and encouraging hazardous weather events. The idea today that humans are responsible for irreversible climate change is generally accepted. In fact, the most recent study by the IPCC states that climate change is already happening and that human activity is the key cause of it. The spectrum of shifts in the atmosphere is leading to many concerns about human livelihoods and ecosystem sustainability. Throughout the history of the world, the earth's atmosphere has altered several times, with occurrences ranging from ice ages to long warmth cycles. Because of human activities associated with CO_2 and other greenhouse gas emissions, the composition of the atmosphere has caused warming and thus affected the climate of the planet.

3.1.2 CARBON DIOXIDE (CO_2) EMISSIONS

Greenhouse gases principally comprise CO_2 emissions (mainly from fossil fuel combustion and forest burning), along with other heat-trapping gases such as methane (from irrigated agriculture, animal husbandry, and oil extraction), nitrous oxide, and various anthropogenic halocarbons. CO_2 is the most significant greenhouse gas contributing to global warming. It is a desirable indicator for gas-controlled combustion units, as it provides a reliable measure of complete combustion (Benemann, 2003; Skjanes et al., 2007). At an average atmospheric concentration of 330 ppm, CO_2 is generally harmless to humans because it does not cause harmful chemical reactions in the body. In a CO_2-rich environment, plants and algae also exhibit enhanced growth (Bhaya et al., 2000). Anthropogenic CO_2 emissions released into the atmosphere from the carbon cycle are estimated at approximately 9 gigatons (Gt) per year. Of these, fossil fuels account for 7.6 Gt, and 1.4 Gt are derived from land use change. Around 55% of this carbon is absorbed by natural processes, and up to 4 Gt accumulates in the atmosphere annually (Lal, 2008).

There are five main carbon reservoirs on Earth: fossil fuel reserves, soils, the atmosphere, the oceans, and terrestrial ecosystems with vegetation (Houghton, 2007). Coal, natural gas, and oil contribute an estimated 5000–10000 Pg C to the first reservoir. Terrestrial ecosystems store around 550 ± 100 Pg C, while soil organic matter accounts for approximately 1500–2000 Pg C in the first meter and about 2300 Pg C in the top 3 m layer (Jobbágy and Jackson, 2000). At the current atmospheric CO_2 concentration (400 ppm), the atmosphere contains nearly 850 Pg C. The oceans represent the largest carbon reservoir, with a cumulative carbon content of approximately 38,000 Pg C, about 50 times more carbon than the atmosphere. Most carbon in the world's oceans is found in intermediate and deep waters, with only 700–1000 Pg C in direct contact with the atmosphere in surface waters (Houghton, 2007). Thus, a large proportion of the planet's CO_2 (approximately 90%) is stored in the oceans. Each year, the oceans absorb about 40% of the anthropogenic CO_2 emitted from fossil fuel combustion and cement production, thereby mitigating the effects of climate change. Although the ocean is considered the largest reservoir of CO_2, the rate of increase in oceanic carbon storage does not appear to keep pace with the current rate of growth in CO_2 emissions. This is because the transport of CO_2 to deeper layers is two to three orders of magnitude slower than atmospheric uptake, and this process is the primary mechanism regulating large-scale CO_2 uptake (Sabine and Tanhua, 2010). Nearly 30% of anthropogenic CO_2 is detected at depths less than 200 m, and around 50% at depths greater than 400 m (Sabine et al., 2004). Several mechanisms are responsible for the transfer of CO_2 from surface waters to deeper layers, collectively known as carbon pumps. The first is the solubility pump, the second is the carbonate counter pump (also known as the alkalinity pump), and the third is the biological pump.

3.1.3 LINKS BETWEEN ALGAE AND CLIMATE CHANGE

Ocean mitigates the environmental effects of global warming by removing heat and carbon from the atmosphere. Owing to human activities, both the physical and chemical properties of the ocean have changed over the past decades, resulting in changes in the composition and functioning of the marine environment and altering the cycling of carbon and nutrients on the surface of the ocean with probable input on the climate system (Riebesell et al., 2009). The increase in oceanic temperature will decrease CO_2 solubility and anthropogenic carbon uptake by 9–15%. As a result, ocean warming would adversely affect primary marine producers. The impact of the rise in ocean warming is quite challenging due to the broad direct and indirect effects and the different ecological relationships of various organisms and functional groups (Behrenfeld, 2011). Higher temperatures would stimulate photosynthetic activity to increase primary productivity in nutrient-free and heterotrophic metabolism regions, intensifying top-down phytoplankton assembly control and decreasing primary productivity (Winder and Sommer, 2012). Likewise, the sudden growth of anthropogenic CO_2 in the ocean would have physical and biological implications. Biologically, the effects of the growing concentration of CO_2 will differ among organisms. The increasing amount of the ocean's aqueous CO_2 concentration reduces pH and thus modifies the balance among the various carbon species in the water. Ocean acidification (OA) will have detrimental effects on calcifying organisms. Ocean carbonation can benefit certain groups of photosynthetic species, particularly those operating a relatively inefficient CO_2 acquisition pathway. However, certain marine algae have biochemical resources in their cells to concentrate CO_2; thus, higher carbon availability does not have a significant impact on productivity (Kroeker et al., 2010). An increase in CO_2 levels may promote the establishment of certain natural phytoplankton assemblages, such as cyanobacteria, while decreasing the competitiveness of others (e.g., some calcifiers). Moreover, the responses of both phytoplankton and macroalgae to elevated CO_2 appear to be strongly influenced by light intensity (Gao et al., 2012; Hepburn et al., 2011). Understanding synergistic and non-synergistic interactions, and how multiple stressors affect ecosystems and natural communities, remains a major challenge for researchers (Darling et al., 2010). The broad recognition of interactions among physical and biological factors, derived from studies examining the impacts of single climate stressors on algal assemblage efficiency, likely requires reexamination. Recent literature has reported contrasting results regarding global trends in ocean primary productivity—positive in some studies (Chavez et al., 2011) and negative in others (Behrenfeld et al., 2006; Boyce et al., 2010)—highlighting the uncertainty surrounding climate change and its potential consequences.

3.1.4 STRATEGIES TO REDUCE THE IMPACT OF CLIMATE CHANGE

Global warming leads to rising sea levels, and various climate anomalies are associated with it, including floods and droughts (Gutierrez et al., 2008; Pedroni et al., 2004; Wang et al., 2008). Considering these significant impacts, it is necessary to implement strategic actions aimed at reducing atmospheric CO_2 concentrations. It is predicted that by 2050, climate change will have substantial implications, including rising temperatures, globally increased rainfall, and elevated CO_2 levels due to greenhouse and agricultural gas emissions (Bengtsson et al., 2009). Various approaches have been proposed worldwide to address the effects of climate change in the coming years. Three major strategies that can lower the atmospheric carbon pool below threshold levels include:

- A global reduction in CO_2 emissions;
- Carbon capture and storage through oceanic, geological, and chemical processes; and
- Exploration of alternative energy sources to replace carbon-based fuels.

Although major carbon capture approaches include oceanic, geological, and chemical transformation methods, biological carbon sequestration appears more promising because of its relative ease, cost-effectiveness, and lower environmental impacts (Jana, 2019).

3.2 ALGAE AS PHOTOSYNTHETIC ORGANISMS AND ABSORBER

Microalgae are aerobic photosynthetic organisms suitable for enhanced CO_2 mitigation or biological carbon fixation. The cultivation of microalgae as a method of energy production, integrated with petroleum-based power plants or wastewater treatment (WWT), could be key to achieving an economically viable process (Kumar et al., 2010). Industrial exhaust gases contain 10–20% CO_2 (Ho et al., 2011) and are considered major contributors to excess CO_2 emissions (Brown and Zeiler, 1993). In response to the energy crisis, the use of microalgae for CO_2 utilization and fuel conversion was introduced (Benemann, 1997; Oswald and Golueke, 1960). Microalgae are capable of producing up to 30 times more oil per unit area than terrestrial oilseed crops due to their high growth rates, often doubling their biomass every 4–6 hours, and their tolerance to varying environmental conditions (Rawat et al., 2013). Depending on the selected algal strain and nutrient conditions, oil content can exceed 80%. For every 1 kg of dry biomass generated, microalgal growth utilizes approximately 1.83 kg of CO_2 (Chisti, 2007). However, production costs remain high (Hughes et al., 2012), and one of the current research challenges is reducing costs, particularly for energy-intensive processes such as harvesting and dewatering (Singh and Ahluwalia, 2013).

Although free-living marine algae account for less than 1% of the Earth's photosynthetic biomass, they contribute approximately 50% of global biogenic carbon fixation (Falkowski, 2012). This is possible because phytoplankton biomass is replaced on average every two to six days worldwide (Behrenfeld, 2011; Falkowski et al., 1998), whereas terrestrial biomass turnover occurs over approximately 19 years. About 45–50 Pg C per year is assimilated by the global phytoplankton community (Falkowski, 2012), making them as critical to the Earth's carbon and CO_2 cycles as all terrestrial plants combined. Ocean productivity per unit area is approximately 140 g C m^{-2} $year^{-1}$, about one-third of that of terrestrial plants (Field et al., 1998). Marine macroalgae offer an attractive alternative to terrestrial crops. Algal production can occur on brownfield land or in the open ocean without competing for arable land. Furthermore, algal farms can be established in waters near industrial centers and power plants that emit CO_2, resulting in additional benefits. The global net primary productivity of cropland is approximately 470 g C m^{-2} $year^{-1}$ (Field et al., 2008), whereas estimates of seaweed carbon capture during photosynthesis are approximately 1600 g C m^{-2} $year^{-1}$ (Duarte et al., 2005).

The net primary production of light and nutrients in the ocean is highly regulated. In oligotrophic oceanic waters, tiny phytoplankton cells dominate the biomass of phytoplankton called picophytoplankton, that is, cyanobacteria and nano-flagellates, with sizes that range from 1–5 μm (Falkowski et al., 1998). Due to a higher surface area-to-volume ratio and a lower thickness of the diffusion boundary layer, picophytoplankton cells may contribute to a substantial portion of the total phytoplankton biomass and development, benefiting larger cells in nutrient-improved environments (Raven, 1998). In nutrient-rich waters, diatoms are stimulated by large phytoplankton cells when solar energy is sufficient, and the thermocline prevents the sinking of cells below the euphotic region (Falkowski et al., 1998). By temporal decoupling with large zooplankton life cycles, these larger cells can expand rapidly and avoid grazing. Such decoupling facilitates the burst of blooms that increase vertical fluxes of organic matter, making a major contribution to carbon sequestration. The total planktonic production in the ocean's eutrophic areas is estimated at 9.1 Pg C year^{-1}. Kelp forests are particularly important macroalgal assemblages found on shallow rocky shores in temperate and Arctic regions, dominated by large brown algae. The total standing crop of kelp forests is estimated to range from 0.015 to 0.02 Pg C (de Vooys, 1979), although it may be up to two times greater (Reed and Brzezinski, 2009). The average net primary productivity of kelp forests is approximately 1000 g C m^{-2} year^{-1} (Reed and Brzezinski, 2009), but productivity values may exceed 3000 g C m^{-2} year^{-1}, as reported for *Macrocystis* and *Laminaria* (Gao and McKinley, 1994).

Low-lipid macroalgae may be used to generate either biogas or alcohol-based fuels. Due to the presence of polysaccharides and sugar alcohols—laminarin and mannitol—macroalgae are promising feedstock candidates for conversion to liquid fuels (Bruton et al., 2009; Hughes et al., 2012). Seaweeds containing natural sugars and other carbohydrates can be processed to produce biogas or ethanol through fermentation, including anaerobic digestion (Bruton et al., 2009). Biogas containing approximately 60% methane can be used to generate heat and electricity or can be compressed for use as a transport fuel. Several genera of seaweed, such as *Macrocystis*, *Gracilaria*, *Ulva*, *Sargassum*, *Laminaria*, *Ascophyllum*, and *Cladophora*, have been tested as potential methane sources (Gunaseelan, 1997). *Laminaria* provided an average of 22 m^3 of methane per ton of wet weight, with a conservative yield of 171 GJ ha^{-1} (Hughes et al., 2012). Although the production of biogas from seaweed farming is considered a significant and technically viable commercial opportunity for macroalgae, the associated costs remain high and economically inefficient under current market conditions (Bruton et al., 2009; Gunaseelan, 1997). Large-scale macroalgal ocean afforestation proposes cultivating seaweeds in eutrophic oceans and using seaweed-digesting microbes to concentrate carbon into 40% bio-carbon dioxide and 60% biomethane, compared with 0.04% in air (N'Yeurt et al., 2012). N'Yeurt et al. (2012) estimated that reducing atmospheric CO_2 concentrations by 100 ppm over 50 years would require afforesting approximately 6% of the ocean surface (Table 3.1).

TABLE 3.1 Global Carbon Sequestration by Marine Ecosystems

System	Global Carbon Sequestration (Pmol C year–1)	References
Phytoplankton	0.75–1.1	Buitenhuis et al. (2013); Field et al. (1998); Laufkötter et al. (2015); Laws et al. (2011)
Macroalgae	0.00083–0.022	Charpy-Roubaud and Sournia (1990); Duarte and Cebrián (1996); Duarte et al. (2013b); Krause-Jensen and Duarte (2016); Smith (1981)
Seagrasses	≤0.009	Duarte et al. (2013b)
Salt marshes	≤0.007	Duarte et al. (2013b)
Mangroves	≤0.0021	Duarte et al. (2013b)

3.3 CARBON CAPTURE POTENTIALS OF MICROALGAE

Various efforts have been made by countries to find an ideal solution to tackle the issues of change in the atmosphere, but no single mitigation technology could provide an appropriate solution to the problem. Biological carbon storage systems can, however, be used to offset carbon emissions that would otherwise be emitted into the environment. It is possible to deploy biological processes in a sustainable and renewable way (Papazi et al., 2008). This technology includes three stages: collecting CO_2 from specific emission sources linked to industry and electricity by using certain techniques, transporting CO_2 to an acceptable storage area, and processing the CO_2 to store it away to prevent it from being released into the environment for a long period of time (Haszeldine, 2009; Rochelle, 2009). However, transporting CO_2 from a source of CO_2 to algal cultures for the purpose of microalgae photosynthesis to improve CO_2 capture efficiency and productivity may be a challenge (Beardall and Raven, 2013; Chang and Yang, 2003). As photosynthetic organisms, algae are well suited to ambient CO_2 capture. Growing algae to capture CO_2 in ambient air will eliminate CO_2 and sequester it in biomass form (Zhu et al., 2008). Algae can be used in CCS technology as a biological capture agent because of their relatively high photosynthetic efficiencies compared to various terrestrial plants, and an alternative means of lowering CO_2 emissions into the atmosphere has been suggested (Pedroni et al., 2004). Therefore, a suitable selection of microalgal strains that are tolerant of high CO_2 concentrations and capable of reducing large quantities of CO_2 efficiently is necessary (Table 3.2).

TABLE 3.2 Some Species of Microalgae That Play a Major Role in CO_2 Bio-Mitigation

Microalgae	Temperature (°C)	Maximum CO_2 Concentration (% v/v) Removed	P Volume (g L–1 day–1)	References
Chlorella	26	Air, 2–15	0.099–1.445	Chiu et al. (2008)
Spirulina	30	12	0.22	De Morais and Costa (2007)
Haematococcus pluvialis	20	16–34	0.076	Huntley and Redalje (2007)
Chlorella kessleri	30	18	0.087	De Morais and Costa (2007)
Chlorella vulgaris	25	10	0.105	Yoo et al. (2010)
Chlorella vulgaris	25	Air	0.024–0.040	Scragg et al. (2002)
Botryococcus braunii	25	10	0.027	Yoo et al. (2010)
Scenedesmus	25	10	0.218	Yoo et al. (2010)
Scenedesmus	25	Flue gas	0.203	Yoo et al. (2010)
Scenedesmus obliquus	30	18	0.14	De Morais and Costa (2007)

3.4 ALGAL PRODUCTION OF BIOFUELS

The ongoing use of fossil fuels is not renewable, and one fifth of the world's CO_2 emissions come from the transport industry (Rawat et al., 2013). The development of biofuels from biological carbon sequestration has emerged as an efficient alternative (Demirbas, 2007; Singh et al., 2011) since the use of sustainable biomass as an energy resource has a positive environmental impact due to lower toxicity and emission profiles. Macroalgae are scarce in lignin-type materials, which during biofuel development are resistant to conversion and increase the cost of the process. The development of fuel where high lipid content of microalgae biomass can be extracted and re-certified into fuel is one of the important uses of microalgae.

Numerous attempts have been made to produce large quantities of commercially viable organic chemicals using algae for use either as liquid or gaseous fuels or, more broadly, as chemical feedstocks (Borowitzka, 2013a). Algae-based biofuels are not as commercially competitive with fuels derived from other fossil sources (Greenwell et al., 2010; Leite et al., 2013; Scott et al., 2010). High-value dietary supplements with low volumes, such as astaxanthin and β-carotene, can be produced economically from microalgal mass cultures, but no data is available on the global development of algae as dietary supplements (Borowitzka, 2013b; Wells et al., 2017). Among a variety of other technical challenges, the economic viability of algal growth requires a selection of acceptable strains and genetic modification of microalgae that are most suitable for mass cultivation for the development of biofuels (Raven and Ralph, 2015; Williams and Laurens, 2010), as well as addressing the problems with the provision of light and CO_2 to mass cultures (Beardall and Raven, 2013; Kenny and Flynn, 2016; Ritchie and Larkum, 2012; Singh and Olsen, 2011). Researchers worldwide are engaged in this emerging field of innovative and cutting-edge research to explore the best suitable strains of algae and thereby develop the most efficient farming practices under different environmental conditions for biofuel production (Douskova et al., 2009; Janssen et al., 2003).

3.5 ENHANCING CO_2 REMOVAL EFFICIENCY IN OPEN AND CLOSED PHOTOBIOREACTORS

The construction of efficient photobioreactors (PBRs) is one of the most critical steps towards a successful process of microalgal CO_2 fixation. Microalgae are grown in open raceway ponds (open system) or closed PBR systems, which are exposed to air or aerated to allow microalgal growth. Open systems are the closest to the natural environment, but closed PBRs are expensive, more regulated, and can control cultivation conditions with a low risk of contamination (Rawat et al., 2013). Concerning CO_2 mitigation using microalgae, open ponds are more favorable than PBRs. In order to optimize biomass productivity, several enclosed designs have been built for PBRs, such as tubular, vertical-column, flat panel, and internally illuminated systems (Chisti, 2007; Pulz, 2001; Tredici and Materassi, 1992;

Ugwu et al., 2008). The most widely applied designs for use in algal PBRs are tubular and bubble column types.

Algal productivity is estimated in terms of biomass produced per day per unit of available surface area (g dry biomass m^{-2} day^{-1}). Open ponds are generally suitable for multi-strain cultivation, whereas PBRs facilitate single-strain cultivation. Productivity levels in open raceway systems range from 5–50 g dry cell weight (DCW) m^{-2} day^{-1}, whereas in PBRs, they range from 27–62 g DCW m^{-2} day^{-1} (Rawat et al., 2013). Suitable species for cultivation in open ponds include *Chlorella*, *Dunaliella salina*, and *Spirulina* (Borowitzka, 1999). It is essential to select optimal microalgal species with high CO_2 fixation efficiency and the ability to produce substantial amounts of valuable products (Ono and Cuello, 2006) to develop effective CO_2 emission mitigation processes. *Chlorella*, *Scenedesmus*, *Dunaliella*, *Spirulina*, and *Chlamydomonas* contain high levels of starch and glycogen, which are useful as raw materials for alcohol production. *Scenedesmus* has been reported to convert approximately 15–25% of atmospheric CO_2 into biodiesel suitable for transport fuel (Ho et al., 2010). For open pond production systems, algae must outcompete wild strains and be robust enough to withstand shear stress in PBR cultivation. High CO_2 tolerance and good settling capability are also desirable characteristics of an ideal algal strain for power plant flue gas remediation (Singh and Ahluwalia, 2013). To achieve favorable economic conditions, hybrid systems combining open ponds and PBRs have been developed. In enclosed PBRs, the initial growth stage is carried out to prevent contamination, followed by a second stage in open ponds. These hybrid systems can produce as much as 20–30 tons ha^{-1} of lipids annually, depending on climatic conditions (Rodolfi et al., 2009). In summary, atmospheric CO_2 levels can be reduced by biomass in two ways—as a carbon store and as an energy source. Primary marine producers (phytoplankton, seaweeds, and seagrasses) are generally more efficient carbon-sequestering agents than their terrestrial counterparts. Moreover, the utilization of anthropogenic CO_2 for algal biomass production as an industrial by-product can contribute to meeting the global demand for food, fodder, and fuel.

3.6 ALGAL BIOENGINEERING TO LIMIT GLOBAL ENVIRONMENTAL CHANGE

Globally, the increase of CO_2 concentrations has a strong influence on photosynthetic algae and other biota (Boyd and Doney, 2002; Cornwall et al., 2017; Doney et al., 2009; Raven and Beardall, 2014; Raven et al., 2017). In fact, this raises the global temperature of the surface of the atmosphere, land, ocean and progressively in the deep ocean. Anthropogenic global environmental change is due to the use of fossil fuels and land use change, leading to increased atmospheric CO_2, which can be eliminated by biological land removal, physical and biological ocean removal (Lenton and Vaughan, 2009; Vaughan and Lenton, 2011). The direct effects of higher temperature on species are exacerbated by the indirect effects of increased temperature, such as the availability of phytoplankton nutrients and the altered mixing depth of the upper layer of the ocean with effects of the incident PAR and UV events (Boyd and Doney, 2002; Raven and Beardall, 2014; Raven et al., 2017; Steinacher et al., 2010).

Bio-geoengineering is a subset of geoengineering aimed at reducing the global rise in temperature, as well as atmospheric and oceanic CO_2 levels. There are primarily two geoengineering categories: carbon dioxide removal (CDR) and solar radiation management (SRM) (Anselm and Hansson, 2014; Lenton and Vaughan, 2009; Raven, 2017; Vaughan and Lenton, 2011). These two approaches may decrease radiative forcing and delay the rise in global temperature. CDR is responsible for slowing the increase in atmospheric CO_2 and its accumulation in water bodies. Chemical sequestration of CO_2 from industrial point sources, or less readily from the atmosphere, may involve non-biological CDR methods. Non-biological SRM involves dispersing reflective particles in the upper atmosphere or increasing the albedo of parts of the Earth's surface (Lenton and Vaughan, 2009; Raven, 2017; Vaughan and Lenton, 2011). Algal bioengineering, through modification of natural processes, may serve as a biological alternative to non-biological geoengineering approaches.

3.7 CONCLUSION

This chapter discusses the potential roles of algae in limiting the rise in atmospheric CO_2 concentrations and global temperatures. Algae play an indispensable role in removing CO_2 through sequestration and the assimilation of inorganic carbon into organic matter from the atmosphere and the photic zone of natural waters. Microalgae cultivation has gained widespread attention in recent years as a viable option for anthropogenic CO_2 conversion because of its potential as a feasible technology for CO_2 sequestration. One promising approach to combating climate change is algae-based CO_2 sequestration and the recovery and storage of carbon released from the global energy system. The production of algae as a fuel source for transport is one potential contribution to limiting temperature rise. Another important role of algae is enhancing current CO_2 sequestration as organic carbon, thereby decreasing the rate of increase in atmospheric CO_2 and partially alleviating ocean acidification. Open-ocean blooms represent a cost-effective method of algal cultivation. PBRs can also produce algae in large quantities; however, their construction and operation are costly. Algal biomass produced through these methods can potentially be used as biofuel. Thus, bio-geoengineering of algae may help reduce the rise in atmospheric and oceanic CO_2 levels and temperatures. Although algae-based carbon capture technology faces several challenges, overcoming these obstacles could not only contribute to mitigating global warming but also reduce dependence on fossil fuels.

KEYWORDS

- **aerobic photosynthetic organisms**
- **biofuel**
- **biomass**
- **bioreactors**
- **carbon emissions**
- **carbon fixation**
- **global warming**
- **macroalgae**
- **photobioreactor**

REFERENCES

Anselm, J., & Hansson, A. (2014). The last chance to save the planet? An analysis of the geoengineering advocacy discourse in the public debate. *Environmental Humanities, 5*, 101–123.

Beardall, J., & Raven, J. A. (2004). The potential effects of global climate change on microalgal photosynthesis, growth and ecology. *Phycological, 43*(1), 26–40.

Behrenfeld, M. J. (2011). Biology: Uncertain future for ocean algae. *Nature Climate Change, 1*(1), 33–34.

Behrenfeld, M. J., O'Malley, R. T., Siegel, D. A., et al. (2006). Climate-driven trends in contemporary ocean productivity. *Nature, 444*(7120), 752–755.

Benemann, J. R. (1997). CO_2 mitigation with microalgae systems. *Energy Conversion and Management, 38*, S475–S479.

Benemann, J. R. (2003). Utilization of carbon dioxide from fossil fuel-burning power plants with biological systems. *Energy Conversion and Management, 34*, 999–1004.

Bengtsson, L., Hodges, K. I., & Keenlyside, N. (2009). Will extratropical storms intensify in a warmer climate? *Journal of Climate, 22*, 2276–2301.

Bhaya, D., Schwarz, R., & Grossman, A. R. (2000). Molecular responses to environmental stress. In B. A. Whitton & M. Potts (Eds.), *The Ecology of Cyanobacteria: Their Diversity in Time and Space* (pp. 397–442). Kluwer Academic.

Borowitzka, M. A. (1999). Commercial production of microalgae: Ponds, tanks, tubes and fermenters. *Journal of Biotechnology, 70*, 313–321.

Borowitzka, M. A. (2013a). Energy from microalgae: A short history. In M. A. Borowitzka & N. R. Moheimani (Eds.), *Algae for Biofuels and Energy* (pp. 1–15). Springer.

Borowitzka, M. A. (2013b). High-value products from microalgae: Their development and commercialization. *Journal of Applied Phycology, 23*, 745–756.

Boyce, D. G., Lewis, M. R., & Worm, B. (2010). Global phytoplankton decline over the past century. *Nature, 466*(7306), 591–596.

Boyd, P. W., & Doney, S. C. (2002). Modeling regional responses by marine pelagic ecosystems to global climate change. *Geophysical Research Letters, 29*(16), 531–534.

Brown, L. M., & Zeiler, K. G. (1993). Aquatic biomass and carbon dioxide trapping. *Energy Conversion and Management, 34*, 1005–1013.

Bruton, T., Lyons, H., Lerat, Y., Stanley, M. S., & Rasmussen, M. B. (2009). *A Review of the Potential of Marine Algae as a Source of Biofuel in Ireland.* Sustainable Energy Ireland.

Buitenhuis, E. T., Hashioka, T., & Le Quéré, C. (2013). Combined constraint on global ocean primary production using observations and models. *Global Biogeochemical Cycles, 27*, 847–858.

Chang, E. H., & Yang, S. S. (2003). Some characteristics of microalgae isolated in Taiwan for bio fixation of carbon dioxide. *Botanical Bulletin of Academia Sinica, 44*, 43–52.

Charpy-Roubaud, C., & Sournia, A. (1990). The comparative estimation of phytoplankton, microphytobenthic and microphytobenthic primary production in the oceans. *Marine Microbial Food Webs, 4*, 31–57.

Chavez, F. P., Messie, M., & Pennington, J. T. (2011). Marine primary production in relation to climate variability and change. *Annual Review of Marine Science, 3*(1), 227–260.

Chisti, Y. (2007). Biodiesel from microalgae. *Biotechnology Advances, 25*(3), 294–306.

Chiu, S. Y., Kao, C. Y., Chen, C. H., Kuan, T. C., Ong, S. C., & Lin, C. H. (2008). Reduction of CO_2 by a high-density culture of *Chlorella sp.* in a semicontinuous photobioreactor. *Bioresource Technology, 99*(9), 3389–3396.

Cornwall, C. E., Revill, A. T., Hall-Spencer, J. M., Milazzo, M., Raven, J. A., & Hurd, C. L. (2017). Inorganic carbon physiology underpins macroalgal responses to elevated CO_2. *Scientific Reports, 7*, 46297.

Darling, E. S., McClanahan, T. R., & Cote, I. M. (2010). Combined effects of two stressors on Kenyan coral reefs are additive or antagonistic, not synergistic. *Conservation Letters, 3*(2), 122–130.

De Mirbas, A. (2007). Alternatives to petroleum diesel fuel. *Energy Sources Part B: Economics, Planning and Policy, 2*(4), 343–351.

De Morais, M. G., & Costa, J. A. V. (2007). Bio fixation of carbon dioxide by *Spirulina sp.* and *Scenedesmus obliquus* cultivated in a three-stage serial tubular photobioreactor. *Journal of Biotechnology, 129*, 439–445.

De Vooys, C. G. N. (1979). Primary production in aquatic environments. In B. Bolin, E. T. Degens, S. Kempe, & P. Ketner (Eds.), *The Global Carbon Cycle* (Scientific Committee on Problems of the Environment (SCOPE) Series 13, p. 491). John Wiley and Sons.

Doney, S. C., Fabry, S. C., Feely, R. A., & Kleypas, J. A. (2009). Ocean acidification: The other CO_2 problem. *Annual Review of Marine Science, 1*, 169–192.

Douskova, I., Doucha, J., Livansky, K., et al. (2009). Simultaneous flue gas bioremediation and reduction of microalgal biomass production costs. *Applied Microbiology and Biotechnology, 82*(1), 179–185.

Duarte, C. M., & Cebrian, J. (1996). The fate of marine autotrophic production. *Limnology and Oceanography, 41*, 1758–1766.

Duarte, C. M., Losada, I. J., Hendriks, I., Mazarrasa, I., & Marba, N. (2013b). The role of coastal plant communities for climate change mitigation and adaptation. *Nature Climate Change, 3*, 961–968.

Duarte, C. M., Middelburg, J. J., & Caraco, N. (2005). Major role of marine vegetation on the oceanic carbon cycle. *Bio Geosciences, 2*(1), 1–8.

Falkowski, P. (2012). Ocean science: The power of plankton. *Nature, 483*(7387), S17–S20.

Falkowski, P. G., Barber, R. T., & Smetacek, V. (1998). Biogeochemical controls and feedback on ocean primary production. *Science, 281*(5374), 200–206.

Field, C. B., Behrenfeld, M. J., Randerson, J. T., & Falkowski, P. (1998). Primary production of the biosphere: Integrating terrestrial and oceanic components. *Science, 281*, 237–240.

Field, C. B., Campbell, J. E., & Lobell, D. B. (2008). Biomass energy: The scale of the potential resource. *Trends in Ecology and Evolution, 23*(2), 65–72.

Gao, K., & McKinley, K. R. (1994). Use of macroalgae for marine biomass production and CO_2 remediation: A review. *Journal of Applied Phycology, 6*(1), 45–60.

Gao, K., Xu, J., Gao, G., Li, Y., Hutchins, D. A., Huang, B., et al. (2012). Rising CO_2 and increased light exposure synergistically reduce marine primary productivity. *Nature Climate Change, 2*, 519–523.

Greenwell, H. C., Laurens, L. M. L., Shields, R. J., Lovitt, R. W., & Flynn, K. J. (2010). Placing microalgae on the biofuel's priority list: A review of the technological challenges. *Journal of the Royal Society Interface, 46*, 703–726.

Gunaseelan, V. N. (1997). Anaerobic digestion of biomass for methane production: A review. *Biomass and Bioenergy, 13*(1, 2), 83–114.

Gutierrez, R., Gutierrez-Sanchez, R., & Nafidi, A. (2008). Trend analysis using nonhomogeneous stochastic diffusion processes: Emission of CO_2; Kyoto protocol in Spain. *Stochastic Environmental Research and Risk Assessment, 22*, 57–66.

Hallock, P. (2005). Global change and modern coral reefs: New opportunities to understand shallow-water carbonate depositional processes. *Sedimentary Geology, 175*, 19–33.

Haszeldine, S. R. (2009). Carbon capture and storage: How green can black be? *Science, 32*, 1647–1651.

Hepburn, C. D., Pritchard, D. W., Cornwall, C. E., et al. (2011). Diversity of carbon use strategies in a kelp forest community: Implications for a high CO_2 Ocean. *Global Change Biology, 17*(7), 2488–2497.

Ho, S. H., Chen, C. Y., Lee, D. J., & Chang, J. S. (2011). Perspectives on microalgal CO2 emission mitigation systems: A review. *Biotechnology Advances, 29*(2), 189–198.

Ho, S. H., Chen, W. M., & Chang, J. S. (2010). *Scenedesmus obliquus* CNW-N as a potential candidate for CO_2 mitigation and biodiesel production. *Bioresource Technology, 101*(22), 8725–8730.

Houghton, J. (2009). *Global Warming: The Complete Briefing* (4th ed.). Cambridge University Press.

Houghton, R. A. (2007). Balancing the global carbon budget. *Annual Review of Earth and Planetary Sciences, 35*, 313–347.

Hughes, A. D., Kelly, M. S., Black, K. D., & Stanley, M. S. (2012). Biogas from macroalgae: Is it time to revisit the idea? *Biotechnology for Biofuels, 5*, 86.

Huntley, M. E., & Redalje, D. G. (2007). CO_2 mitigation and renewable oil from photosynthetic microbes, in a new appraisal. *Mitigation and Adaptation Strategies for Global Change, 12*, 573–608.

Jana, B. B. (2019). CO_2 mitigation potentials of microalgae: Its expansion to a new dimension for closing the loop of carbon between source and acquisition through food chain of phytophagous fish in open ponds. *Advances in Biotechnology and Microbiology, 12*(1), 003–008.

Janssen, M., Tramper, J., Mur, L. R., & Wijffels, R. H. (2003). Enclosed outdoor photobioreactors: Light regime, photosynthetic efficiency, scale-up, and future prospects. *Biotechnology and Bioengineering, 81*(2), 193–210.

Jobbagy, E. G., & Jackson, R. B. (2000). The vertical distribution of soil organic carbon and its relation to climate and vegetation. *Ecological Applications, 10*(2), 423–436.

Kenny, P., & Flynn, K. J. (2016). Coupling a simple irradiance description to a mechanistic growth model to predict algal production in industrial-scale solar-powered photobioreactors. *Journal of Applied Phycology, 28*, 3203–3312.

Krause-Jensen, D., & Duarte, C. M. (2016). Substantial role of macroalgae in marine carbon sequestration. *Nature Geoscience, 9*, 737–742.

Kroeker, K. J., Kordas, R. L., Crim, R. N., & Singh, G. G. (2010). Meta-analysis reveals negative yet variable effects of ocean acidification on marine organisms. *Ecology Letters, 13*(11), 1419–1434.

Kumar, A., Ergas, S., Yuan, X., Sahu, A., Zhang, Q., Dewulf, J., et al. (2010). Enhanced CO_2 fixation and biofuel production via microalgae: Recent developments and future directions. *Trends in Biotechnology, 28*(7), 371–380.

Lal, R. (2008). Sequestration of atmospheric CO_2 in global carbon pools. *Energy and Environmental Science, 1*, 86–100.

Laufkötter, C., Vogt, M., Gruber, N., Aita-Noguchi, M., Aumont, O., Bopp, L., et al. (2015). Drivers and uncertainties of future global marine primary production in marine ecosystem models. *Bio Geosciences, 12*, 6955–7984.

Laws, E. A., D'Sa, E., & Naik, P. (2011). Simple equations to estimate ratios of new and export production to total production from satellite-derived estimates of sea surface temperature and primary production. *Limnology and Oceanography Methods, 9*, 593–601.

Leite, G. B., Adeelaziz, E. M., & Hallenbeck, P. C. (2013). Algal biofuels: Challenges and opportunities. *Bioresource Technology, 145*, 134–141.

Lenton, T. M., & Vaughan, N. E. (2009). The radiative forcing potential of different climate geoengineering options. *Atmospheric Chemistry and Physics, 9*, 5539–5561.

McMichael, A. J. (2001). *Human Frontiers, Environments and Disease: Past Patterns, Uncertain Futures*. Cambridge University Press.

N'Yeurt, A. D., Chynoweth, D. P., Capron, M. E., Stewart, J. M., & Hasan, M. A. (2012). Negative carbon via ocean afforestation. *Process Safety and Environmental Protection, 90*(6), 467–474.

Ono, E., & Cuello, J. (2006). Feasibility assessment of microalgal carbon dioxide sequestration technology with photobioreactor and solar collector. *Biosystems Engineering, 95*(4), 597–606.

Oswald, W. J., & Golueke, C. G. (1960). The biological conversion of solar energy. *Advances in Applied Microbiology, 11*, 223–242.

Papazi, A., Makridis, P., Divanach, P., & Kotzabasis, K. (2008). Bioenergetic changes in the microalgal photosynthetic apparatus by extremely high CO_2 concentrations induce an intense biomass production. *Physiologiae Plantarum, 132*, 338–349.

Patz, J. A., McGeehin, M. A., Bernard, S. M., Ebi, K. L., Epstein, P. R., Grambsch, A., et al. (2000). The potential health impacts of climate variability and change for the United States: Executive summary of the report of the health sector of the U.S. National Assessment. *Environmental Health Perspectives, 108*(4), 367–376.

Pedroni, P., Davison, J., Beckert, H., Bergman, P., & Benemann, J. (2004). A proposal to establish an international network on bio fixation of CO_2 and greenhouse gas abatement with microalgae. *Journal of Energy and Environmental Research, 1*(1), 1–20.

Pulz, O. (2001). Photobioreactors: Production systems for phototrophic microorganisms. *Applied Microbiology and Biotechnology, 57*(3), 287–293.

Raven, J. A. (1998). The twelfth Tansley Lecture. Small is beautiful: The picophytoplankton. *Functional Ecology, 12*, 503–513.

Raven, J. A. (2017). The possible roles of algae in restricting the increase in atmospheric CO_2 and global temperature. *European Journal of Phycology, 52*(4), 506–522.

Raven, J. A., & Beardall, J. (2014). CO_2 concentrating mechanisms and environmental change. *Aquatic Botany, 118*, 23–37.

Raven, J. A., & Ralph, P. J. (2015). Enhanced biofuel production using optimality, pathway modification and waste minimization. *Journal of Applied Phycology, 27*, 1–31.

Raven, J. A., Beardall, J., & Baracaldo, P. S. (2017). The possible evolution and future of CO_2-concentrating mechanisms. *Journal of Experimental Botany, 68*(14), 3701–3716.

Rawat, I., Kumar, R. R., Mutanda, T., & Bux, F. (2013). Biodiesel from microalgae: A critical evaluation from laboratory to large scale production. *Applied Energy, 103*, 444–467.

Reed, D. C., & Brzezinski, M. A. (2009). Kelp forests. In D. Laffoley & G. Grimsditch (Eds.), *The Management of Natural Coastal Carbon Sinks* (pp. 31–38). IUCN.

Riebesell, U., Kortzinger, A., & Oschlies, A. (2009). Sensitivities of marine carbon fluxes to ocean change. *Proceedings of the National Academy of Sciences of the United States of America, 106*(49), 20602–20609.

Ritchie, R. J., & Larkum, A. W. D. (2012). Modeling photosynthesis in shallow algal production pools. *Photosynthetic, 50*, 481–500.

Rochelle, G. T. (2009). Amine scrubbing for CO_2 capture. *Science, 325*, 1652–1653.

Rodolfi, L., Chini Zittelli, G., Bassi, N., Padovani, G., Biondi, N., Bonini, G., et al. (2009). Microalgae for oil: Strain selection, induction of lipid synthesis and outdoor mass cultivation in a low-cost photobioreactor. *Biotechnology and Bioengineering, 102*(1), 100–112.

Sabine, C. L., & Tanhua, T. (2010). Estimation of anthropogenic CO_2 inventories in the ocean. *Annual Review of Marine Science, 2*(1), 175–198.

Sabine, C. L., Feely, R. A., Gruber, N., Key, R. M., Lee, K., Bullister, J. L., et al. (2004). The oceanic sink for anthropogenic CO_2. *Science, 305*(5682), 367–371.

Scott, S. A., Davey, M. P., Dennis, J. S., Horst, I., Howe, C. J., Lea-Smith, D. J., et al. (2010). Biodiesel from algae: Challenges and prospects. *Current Opinion in Biotechnology, 21*, 277–286.

Scragg, A. H., Illman, A. M., Carden, A., & Shales, S. W. (2002). Growth of microalgae with increased caloric values in a tubular bioreactor. *Biomass and Bioenergy, 23*, 67–73.

Singh, A., & Olsen, S. I. (2011). Critical review of biochemical conversion, sustainability and life cycle assessment of algal biofuels. *Applied Energy, 28*, 3548–3555.

Singh, A., Nigam, P. S., & Murphy, J. D. (2011). Renewable fuels from algae: An answer to debatable land-based fuels. *Bioresource Technology, 102*(1), 10–16.

Singh, U. B., & Ahluwalia, A. S. (2013). Microalgae: A promising tool for carbon sequestration. *Mitigation and Adaptation Strategies for Global Change, 18*(1), 73–95.

Skjanes, K., Lindblad, P., & Muller, J. (2007). BioCO_2—A multidisciplinary, biological approach using solar energy to capture CO_2 while producing H_2 and high value products. *Biomolecular Engineering, 24*, 405–413.

Smith, S. V. (1981). Marine macrophytes as a global carbon sink. *Science, 211*, 838–840.

Steinacher, M., Joos, F., Frölicher, T. L., Bopp, L., Cadule, P., Cocco, V., et al. (2010). Projected 21st century decrease in marine productivity: A multi-model analysis. *Bio Geosciences, 7*, 979–1005.

Tredici, M. R., & Materassi, R. (1992). From open ponds to vertical alveolar panels—The Italian experience in the development of reactors for the mass cultivation of phototrophic microorganisms. *Journal of Applied Phycology, 4*(3), 221–231.

Treydte, K. S., Schleser, G. H., Helle, G., Frank, D. C., Winiger, M., Haug, G. H., et al. (2006). The twentieth century was the wettest period in Northern Pakistan over the past millennium. *Nature, 440*, 1179–1182.

Ugwu, C. U., Aoyagi, H., & Uchiyama, H. (2008). Photobioreactors for mass cultivation of algae. *Bioresource Technology, 99*(10), 4021–4028.

Vaughan, N. E., & Lenton, T. M. (2011). A review of climate geoengineering proposals. *Climatic Change, 109*, 745–790.

Walther, G. R., Post, E., Convey, P., Menzel, A., Parmesan, C., Beebee, T. J. C., et al. (2002). Ecological responses to recent climate change. *Nature, 416*, 389–395.

Wang, B., Li, Y., Wu, N., & Lan, C. Q. (2008). CO_2 bio-mitigation using microalgae. *Applied Microbiology and Biotechnology, 79*, 707–718.

Wells, M. L., Potin, P., Craigie, J. S., Raven, J. A., Merchant, S. S., Helliwell, K. E., et al. (2017). Algae as nutritional and functional food sources: Revisiting our understanding. *Journal of Applied Phycology, 29*, 949–982.

Williams, P. J. le B., & Laurens, L. M. (2010). Microalgae as biodiesel and biomass feedstocks: Review and analysis of the biochemistry, energetics and economics. *Energy and Environmental Science, 3*, 554–590.
Winder, M., & Sommer, U. (2012). Phytoplankton response to a changing climate. *Hydrobiological, 698*(1), 5–16.
Yoo, C., Jun, S. Y., Lee, J. Y., Ahn, C. Y., & Oh, H. M. (2010). Selection of microalgae for lipid production under high levels carbon dioxide. *Journal of Bioresource Technology, 101*, S71–S74.
Zhu, X. G., Long, S. P., & von Caemmerer, S. (2008). What is the maximum efficiency with which photosynthesis can convert solar energy into biomass? *Current Opinion in Biotechnology, 19*, 153–159.

CHAPTER 4

Microalgae for the Prevention of Organic and Inorganic Pollution in the Environment

TIELE MEDIANEIRA RIZZETTI,[1] MAIARA PRISCILLA DE SOUZA,[1] GLEISON DE SOUZA CELENTE,[1] GUSTAVO STOLZENBERG COLARES,[2] ANDREA MARISA SANCHEZ-BARRIOS,[1] MARCONDES MAFACIOLLI PACHECO,[1] GISLAYNE ALVES OLIVEIRA,[3] LISIANNE BRITTES BENITEZ,[1] ENIO LEANDRO MACHADO,[2] LÚCIA HELENA RIBEIRO RODRIGUES,[3] and ROSANA DE CASSIA DE SOUZA SCHNEIDER[1,*]

[1]*Environmental Technology Postgraduate Program and Center of Excellence in Oilchemistry and Biotechnology, University of Santa Cruz do Sul, Santa Cruz do Sul, Rio Grande do Sul, Brazil*

[2]*Environmental Technology Postgraduate Program, University of Santa Cruz do Sul, Santa Cruz do Sul, Rio Grande do Sul, Brazil*

[3]*Postgraduate Program in Water Resources and Environmental Sanitation, Federal University of Rio Grande do Sul, Porto Alegre, Rio Grande do Sul, Brazil*

*Corresponding author

ABSTRACT

A diversity of contaminants, such as heavy metals and emerging contaminants, has been observed in wastewater. To prevent environmental

Environmental Phycology: Climate Change Mitigation and Hazardous Pollutant Cleanup Using Algal Resources. Jeyabalan Sangeetha and Devarajan Thangadurai (Eds.)

DOI: 10.1201/9781779647085-4

pollution and meet the demands of a growing global population, advanced wastewater treatment (WWT) methods are required. However, excess nutrients and contaminants may require further treatment. Bioremediation using microalgae can remove contaminants from the environment, and the biomass obtained from this process is rich in valuable bioproducts. Thus, microalgae cultivation combined with other processes offers an alternative treatment approach for pollutant removal along with biomass generation. The removal of pollutants by microalgae involves different mechanisms, and the selection of microalgae and their tolerance to pollutants are essential characteristics for understanding bioremediation. This chapter presents metabolic mechanisms responsible for removing organic and inorganic contaminants using microalgae. Furthermore, this chapter discusses the potential of microalgae-bacteria consortia for wastewater bioremediation, the integration of treatment systems, and the future perspectives and challenges of WWT and biomass valorization.

4.1 INTRODUCTION

The rapid industrial expansion and agricultural pressure to sustain food production for the increasing population lead to severe environmental consequences. A diversity of contaminants has been observed in wastewaters, consisting of high organic load, total nitrogen, total phosphorus, heavy metals, and emerging contaminants, among others. Wastewater treatments (WWTs) with advanced methods are required to meet the growing world population's demands and prevent environmental pollution.

Wastewaters with excess nutrients and contaminants may require further treatment. The use of microalgae for remedying aggressive or unbalanced molecules in the environment requires the development of clean technologies that include stages of bioremediation and the use of the produced biomass. Integrating microalgae cultivation with other processes offers an alternative treatment approach for removing pollutants; it simultaneously allows the conversion of pollutants into biomass and the obtaining of valuable bioproducts (de Souza et al., 2019). The pollutant removal and biomass generation by microalgae can be strongly influenced by wastewater properties and microalgae species (Lu et al., 2020). Generally, removing contaminants by microalgae involves different processes, such as biosorption, bioaccumulation, and biodegradation.

The selection of microalgae for bioremediation has focused on characteristics such as strain, tolerance to the target pollutant, size, and life stage

(Suresh et al., 2015). The genera *Scenedesmus* and *Chlorella* are among the most extensively researched worldwide for microalgae cultivation (Zhuang et al., 2018), highlighting the relationship between microalgal growth and pollutant removal. Microalgae-based bioremediation technology can be used in conjunction with other microorganisms, such as in algae-bacteria consortia. Microalgae produce oxygen via photosynthesis to promote bacterial growth, while utilizing the carbon dioxide (CO_2) produced by bacteria as a carbon substrate (Kumari et al., 2016; Nie et al., 2020).

Considering this context, we present an analysis of metabolic mechanisms involved in contaminant removal by microalgae. In addition, we highlight the microalgal species used, bioremediation processes, microalgae-bacteria consortia, integration of treatment systems, future perspectives and challenges of WWT, and biomass valorization.

4.2 BIOLOGICAL AREAS AND FOCUS FOR MICROALGAE STUDIES

Biological studies aim to understand living organisms' distribution, evolution, structure, function, growth, and many other aspects. When it comes to microalgae, the main focus lies on cultivation. In the last decades, the primary interest in this area has been surrounding which species are capable of removing pollutants from water bodies, understanding their mechanisms, functional combinations with other microbial organisms, and their use in product development from waste (Martínez-Bisbal et al., 2019; Nie et al., 2020; Yadav and Sen, 2018).

Up to date, hundreds of thousands of identified/characterized photosynthetic microalgae species are efficient at removing pollutants. Unidentified species could still influence the well-known ones for ecological adaptability (Asnaghi et al., 2017). Microalgae's capacity to use sunlight to capture atmospheric carbon allows them to grow in environments with limited nutrients. However, this trait makes laboratory analyses challenging due to the complications and limitations involved in mimicking nature's environmental parameters. Consequently, results become inconsistent when applied in natural settings, restricting our understanding of the organisms' bioremediation capacity (Rawat et al., 2013).

Based on this, designing a controlled environment that benefits/enhances their capacity to grow and produce biomass becomes a priority and a challenge (Benedetti et al., 2018; Litchman et al., 2007). Creating proper conditions helps their organelles produce chemical molecules that aid in the "cleaning" of the environment, having an ecological role. A future outcome would be to use them for drug discovery and development while making the process available for many countries that currently find it unsustainable through traditional methods (Agregán et al., 2018; Delalat et al., 2015; Gregory et al., 2013; Taunt et al., 2018).

Microalgae can grow either in open or closed systems, making it a relatively simple approach to target/understand multiple environmental questions related to bioremediation. Since open systems are harder to control and evaluate (the systems approach has been established for almost 50 years now) (Deruyck, 2019; Zhan et al., 2017), closed environments allow us to understand their direct interaction with other microorganisms (bacteria, fungi, and yeast) as well as the mechanisms used for the removal of organic and inorganic compounds. These combinations improve and reduce the cost of the bioremediation process (Hu et al., 2019). Still, bioproducts like biodiesel are considered to be a challenge for scientists in this area (Fazal et al., 2018, Yadav and Sen, 2018), and modifications and constant evaluation of parameters are needed to achieve this goal, focusing on factors that, when modified, alter the natural physiological processes of these organisms (for instance, species, cultivation, harvesting).

Considered to be great alternatives for the use and development of all kinds of products, their use to recover areas classified as inadequate for agriculture gives many countries the alternative for developing and producing their nutritional, health-related, and personal care products (Devadasu et al., 2019). Still, the conjunction of multiple scientific specialties will be needed to ensure their safety and applicability.

Figure 4.1 shows the biological areas and focuses on microalgae studies. The growth and cultivation improvement serve as the starting point and base for removing pollutants and extracting bioproducts from microalgae cells, contributing to sustainable development. It must also evaluate the need to apply molecular and synthetic biology to improve harvesting and product development from microalgae.

The future points toward using these organisms in a molecular approach, specifically through genetic engineering and synthetic biology, to enhance the desirable traits of some of these species (Fu et al., 2019; Nielsen and Keasling, 2016). These approaches, coupled with more intense phenomics studies, could make the selection process of microalgae and bacteria/fungi combinations more consistent and effective (Furbank and Tester, 2011). Reviewing cultivation methods and testing new ways of doing so, plus considering the use of genetics and molecular aspects to increase their potential for bioproducts at an acceptable and profitable level, will be the following objectives to study in the coming years and decades and possibly, a partner for agricultural biotechnology.

In this context, phycoremediation is considered a clean and cost-effective technology for water purification and has been demonstrated to be a suitable approach for efficiently reducing eutrophic loadings (Lu et al., 2020), heavy metals (Leong and Chang, 2020), and emerging contaminants such as synthetic organic compounds (Sutherland and Ralph, 2019), as well as for removing acute toxicity and genotoxicity (de Souza et al., 2020; Silveira et al., 2020). However, several factors can strongly influence microalgal performance in terms of biomass generation and pollutant removal, including wastewater characteristics and microalgal species (Lu et al., 2020), the applied pretreatment (when present), and the cultivation mode (e.g., open ponds or photobioreactors [PBRs]) (Salama et al., 2017).

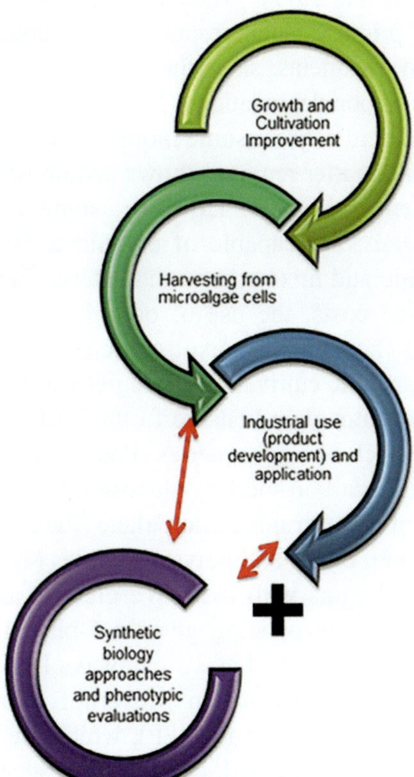

FIGURE 4.1 Key biological areas and focal points for microalgae studies.

4.2.1 MECHANISMS OF ORGANIC COMPOUNDS REMOVAL IN WASTEWATER BY MICROALGAE

Water pollution and resource shortages demand new approaches to mitigate pollution and recover key compounds from wastewater. The organic composition of wastewater varies depending on its source; however, it is often characterized by lipid-like compounds in domestic and food effluents and humic-like compounds in textile effluents and landfill leachates (Rodríguez-Vidal et al., 2020). Heterotrophic and mixotrophic microalgae utilize organic carbon from wastewater as a source of energy and carbon (Murwanashyaka et al., 2020). The composition and pathways associated with compound assimilation and content in microalgae are influenced by carbon concentration (Lin et al., 2017) and the C:N:P ratio (Murwanashyaka et al., 2020). The mechanism of organic carbon assimilation by microalgae remains controversial (Lin et al., 2017). Nevertheless, some researchers identify assimilation as the primary factor responsible for organic carbon removal in microalgal-based treatments, since the ability of microalgae to produce oxygen for organic carbon degradation is limited (Gong et al., 2020).

Microalgae preferentially consume inorganic matter (Tan et al., 2019). Their use for organic matter removal shows certain limitations, especially for highly persistent pollutants. However, some microalgae, such as *Scenedesmus obliquus*, are capable of organic carbon assimilation due to their heterotrophic and mixotrophic characteristics (Shen et al., 2015). Therefore, to reduce costs, the use of organic matter is recommended instead of artificial chemicals in mixotrophic cultures (Tan et al., 2019). Additionally, mixotrophic cultivation strategies can be adopted to achieve significant nutrient removal and enable further utilization of the biomass for other purposes, such as algorefineries (Patel et al., 2020).

Among all organic carbon sources, glucose is the most commonly used for the cultivation of heterotrophic microalgae (Perez-Garcia et al., 2011), and it is converted into oligo- and polysaccharides (Tanner, 2000). Carbohydrate metabolism begins with oxidative glucose assimilation through hexose phosphorylation, producing glucose-6-phosphate (Droop, 1974; Perez-Garcia et al., 2011). Two known pathways lead to aerobic glucose breakdown in microalgae: the Embden–Meyerhof–Parnas (EMP) pathway and the pentose phosphate pathway (PPP), which are the primary glycolytic processes under light and dark conditions, respectively (Hammed et al., 2016; Neilson and Lewin, 1974).

In addition to glucose, glycerol is another substrate used for microalgal growth. The assimilation process involves enzymes such as glycerol kinase, *sn*-glycerol-3-phosphate dehydrogenase, triose phosphate isomerase, and NAD^+-dependent oxidoreductase. Glycerol is converted into glyceraldehyde-3-phosphate and glycerate, which are further used as intermediates in the Embden–Meyerhof–Parnas pathway of glycolysis to form pyruvate, which is subsequently metabolized in the tricarboxylic acid cycle (TCA) for complete glucose oxidation (Neilson and Lewin, 1974; Perez-Garcia et al., 2011).

One of the most common carbon sources in wastewater for microalgal lipid accumulation is acetic acid (acetate), which is commonly present in palm oil mill effluent (POME). Starch breakdown can also result in volatile fatty acids, such as propionic, *iso*-butyric, *n*-butyric, *iso*-valeric, and *n*-valeric acids (Tan et al., 2019). Acetic acid at high concentrations can be harmful to some microalgae, except for *Chlamydomonas mundana*, which can benefit from the presence of acetate. The conversion of acetic acid into lipids is a complex process involving several enzymes for its biotransformation. First, acetic acid is assimilated into coenzyme A (CoA) to form acetyl-CoA and is oxidized in the glyoxylate cycle and the TCA to provide carbon skeletons, energy in the form of ATP, and reducing equivalents such as NADH (Jasni et al., 2020; Perez-Garcia et al., 2011). Subsequently, acetyl-CoA is converted into malonyl-CoA, catalyzed by acetyl-CoA carboxylase and type II fatty acid synthase. Malonyl-ACP is then formed through the transfer of malonyl-CoA to acyl carrier protein (ACP), catalyzed by malonyl-CoA:ACP transacylase (MCAT). Thereafter, it is repeatedly condensed by β-ketoacyl-ACP synthase, reduced by β-ketoacyl-ACP reductase, dehydrated by β-hydroxyacyl-ACP dehydratase, and finally reduced by enoyl-ACP reductase to form saturated fatty acids (Ren et al., 2019). Thus, the availability of acetic acid in the medium strongly influences fatty acid production.

Another nutrient often found in wastewater is nitrogen. Nitrogen is present primarily in domestic effluent as NH_4^+, making nitrification and denitrification the reliable pathways for its removal and readily available to microalgae. Nevertheless, some organic nitrogen may be present in the form of dissolved combined amino acids (DCAA), dissolved free amino acids (DFAA), urea, proteins, and ethylene diamine tetraacetic acid (EDTA) (Xiu et al., 2019). Some microalgae can utilize organic forms in the absence of inorganic nitrogen if the dissolved organic nitrogen (DON) is bioavailable to them. Besides, the available DON can account for up to 80% of the total DON. Sun and Simsek (2017) proved that some species could reach a reduction of up to 58.8% of DON.

Additionally, after NH_4^+, there is no difference in preference among urea, nitrate, and DFAA as nitrogen sources; this shows that as long as the organic nitrogen has a small chain, the microalgae may be able to assimilate it (Xiu et al., 2019). There are two enzymes responsible for urea catalyzing: urea amidolyase and urease. The former represents a reaction among urea, bicarbonate anion, and ATP to form ADP, orthophosphate, and urea-1-carboxylate. The latter consists of the reaction between urea and water molecules to form CO_2 and ammonia (Perez-Garcia et al., 2011).

Wastewater comprises mainly organic carbon and nutrients; some organic contaminants, even in trace concentrations, such as antibiotics, endocrine disruptors, and pesticides, can be present in wastewater. The composition of the cell wall of microalgae facilitates the adsorption of organic contaminants from water. However, microalgae have different removal efficiencies for pollutants, and metabolic kinetics depend on the target compound (Nie et al., 2020). Figure 4.2 demonstrates the general concept of organic compound assimilation and biodegradation by microalgae.

Kiki et al. (2020) evaluated the potential of four microalgae—*Haematococcus pluvialis*, *Selenastrum capricornutum*, *Scenedesmus quadricauda*, and *Chlorella vulgaris*—to dissipate 10 different antibiotics. Their results showed that antibiotic biotransformation occurs through several mechanisms, such as hydroxylation, amination, deamination, oxidation, demethylation, and ring cleavage.

Endocrine-disrupting chemicals (EDCs) are also present in domestic effluents. Their fate involves three intracellular steps: bioadsorption, bioaccumulation within the cell, and intracellular degradation, to which cytochrome P450 enzyme families contribute significantly. Extracellular degradation also plays an essential role, since some metabolites are excreted from the cell and may be degraded externally or reenter the cell through bioadsorption (Xiong et al., 2018). Among the EDCs are estrogens, which microalgae can degrade into various metabolites through different pathways activated depending on the specific estrogen. These pathways may include hydroxylation followed by reduction and methylation reactions. As a result, the metabolites produced exhibit lower estrogenic activity than the parent compounds. Thus, microalgae enable the bioremediation of these compounds prior to discharge (Wang et al., 2019).

Petroleum spills may contaminate water bodies and add alkylated polycyclic aromatic hydrocarbons (APAHs). One of the mechanisms responsible for the removal of APAHs is the direct absorption of algae cells. However, the degradation of these compounds is dependent on the microalgae species. Another pathway for APAH removal is its transformation into hydroxylated APAHs by adding an OH onto methyl groups. This occurs when it enters the intracellular matrix of the microalgae to increase its polarity, and later it is excreted into the extracellular matrix (Luo et al., 2020).

Pesticide removal from the aqueous phase by microalgae involves different processes, including biosorption, bioaccumulation, and biodegradation. Examples of biosorption mechanisms include adsorption, ion exchange, surface complexation, and precipitation (Bilal et al., 2018; Nie et al., 2020). Hussein et al. (2017) investigated the potential of *Chlorella vulgaris* to remove pesticides (namely atrazine, simazine, molinate, isoproturon, carbofuran, propanil, dimethoate, metolachlor, pendimethalin, and pyriproxyfen) from the aqueous phase. These pesticides were removed via biosorption, with removal efficiencies ranging from 87% to 96.5%. Similarly, Weis et al. (2020) assessed the performance of microalgae-induced bioremediation of the pesticide bifenthrin. Their results demonstrated the capability of native microalgae, primarily composed of *Lobochlamys segnis* and *Parachlorella kessleri*, to remove 99% of the pesticide in combination with photodegradation. Ata et al. (2012) identified hydroxyl, carboxyl, and amine groups as the predominant active surface groups involved in the sorption of 2,4-dichlorophenoxyacetic acid (2,4-D) by *Gracilaria verrucosa*. Thus, pesticide sorption efficiency depends on the active surface groups and the properties of the microalgae.

Biodegradation is a necessary process for microbial removal of contaminants from water. The degradation of pesticides is a process that occurs in three steps (Kumar and Singh, 2017; Nie et al., 2020). The first step consists of oxidation, reduction, or hydrolysis reactions, making the pesticide more water-soluble and transforming it into less toxic metabolites. In the second step, the conjugation of a pesticide or the formation of pesticide metabolites with glutathione, amino acids, or sugars occurs. These metabolites may be stored in cellular organelles and generally have little or no toxicity. The third step involves the conversion of the metabolites into secondary nontoxic conjugates. Therefore, the biodegradation mechanism of pesticides depends directly on the nature of the pesticides and microbial mechanisms (Kumar and Singh, 2017; Nie et al., 2020).

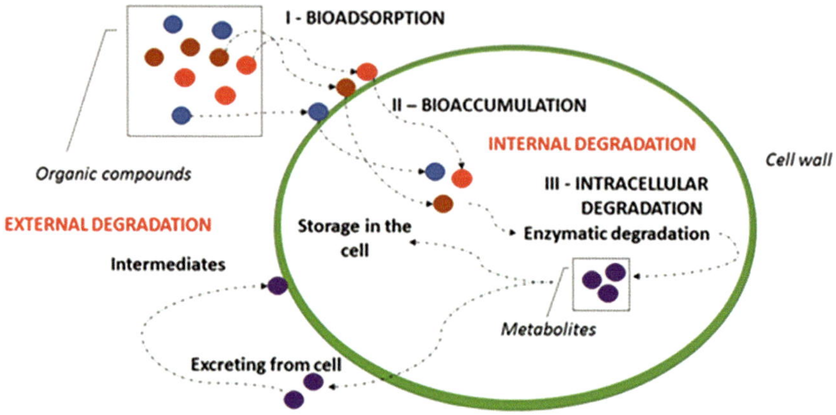

FIGURE 4.2 General concept of the biodegradation of organic pollutants.

Source: Adapted from Nie et al. (2020); Serejo et al. (2020); and Xiong et al. (2018).

4.2.2 MECHANISMS OF INORGANIC COMPOUNDS REMOVAL IN WASTEWATER BY MICROALGAE

Regarding nutrient removal, several mechanisms are involved, and one of the main advantages of microalgae is nutrient recovery through biomass harvesting. The most common removal methods include assimilation and accumulation by microalgae, as well as nitrification and denitrification processes (Molinuevo-Salces et al., 2019). Concerning ammonia nitrogen removal ($NH_3(g) + H_2O(l) \leftrightarrow NH_4^+(aq) + OH^-(aq)$), which is a major water pollutant due to its eutrophication potential and toxic effects, one advantage of using algae is their ability to assimilate ammonia nitrogen into organic nitrogen, thereby reducing its toxicity (Chen and Wang, 2020). Microalgae can assimilate nitrogen under both autotrophic and heterotrophic conditions (Pacheco et al., 2015). Additionally, stripping may be associated with the NH_3 removal mechanism, particularly when pH values above 9.5 are present (de Souza et al., 2019). On the other hand, phosphorus can be removed through microalgal assimilation and chemical precipitation under high alkalinity conditions (Lu et al., 2020). Phosphorus is essential for microalgal growth, as it is required for several metabolic activities, including energy transfer and DNA synthesis. In these systems, phosphorus removal usually co-occurs with nitrogen assimilation; thus, phosphorus can also be recovered from the biomass (Molinuevo-Salces et al., 2019). Phosphorus is generally supplied in excess in wastewater as phosphate salts, which may precipitate with other metal ions in the medium. However, microalgae can assimilate most organic and inorganic forms of phosphorus after mineralization and conversion to phosphates (Nagarajan et al., 2020). Moreover, through photosynthesis, microalgae release oxygen and increase pH levels in water (Bellucci et al., 2020), thereby influencing nutrient and other pollutant removal.

In addition to removing these inorganic compounds, microalgae can capture heavy metals, such as Cr, Pb, Fe, and Co ions, or metallic nanostructures, such as ZnO micro/nano-ferns. This capacity is an essential decontamination process in water (Serrà et al., 2020). Since 1990, several studies have reported microalgae's potential for metal bioremediation due to the ability of metal biosorption and/or bioaccumulation from the environment through mechanisms recognized as adsorption (extracellular) and absorption (intracellular) (Urrutia et al., 2019). The first mechanism is characterized by the rapid superficial interaction of heavy metal ions with functional groups available in the microalgae cell wall. The second mechanism performs the slow internalization of metal ions inside the cell wall (Tripathi et al., 2019).

Bioaccumulation selectivity is better than biosorption; however, the cost is higher and requires more infrastructure. In biosorption, adding biomass to the treatment process is possible without any system modifications prepared for chemical adsorption. Biosorption with dried microalgae biomass can be advantageous since it is reversible. The main parameters to control effective biosorption include temperature, pH, biosorbent dose, contact time, functional groups, solubilized metal concentration, agitation, and size of biosorbent particles (Beni and Esmaeili, 2020).

These removal mechanisms can be optimized in systems that form biofilms (Mantzorou et al., 2018; Miranda et al., 2017). Recent studies have shown that microalgae biofilms in the filter medium or substrate can optimize the potential for heavy metal removal. This is possible considering biofilms offer resistance mechanisms to microorganisms, protecting the biofilm community against shear forces, nutrient deprivation, and other environmental changes, such as pH. Thus, biofilms have properties that make them highly effective in heavy metal removal in polluted water. These processes can be an alternative due to the functional groups found in exopolysaccharides (EPS) present around microorganisms' cell walls, for example, amino, carboxylic, hydroxyl, phospholipids, proteins, and polysaccharides (Bremner et al., 2020; Shen et al., 2020).

According to Leong and Chang (2020), despite the efficiency of heavy metal removal, some aspects are still considered challenges to understanding this process, such as (a) the competition for nutrients by bacteria, fungi, and other microorganisms in wastewater contaminated by heavy metals; (b) nutrient variability. In the clean technology context, researchers should focus on the screening and choice of microalgae strains that are capable of accumulating valuable by-products; reliable CO_2 storage and low need for nutrients; resistance to predatory organisms and robustness to the existence of other species; and the ability to self-flocculate to facilitate the biomass harvesting step with a low cost-benefit ratio. Also, genetic, metabolic, and molecular engineering are necessary to improve microalgae strains' adaptive capacity, specificity, and robustness to remove heavy metals and other inorganic compounds (Leong and Chang, 2020).

4.2.3 MICROALGAE SPECIES USED IN THE REMEDIATION

Microalgae accomplish bioremediation through two mechanisms: a passive mechanism—where a pollutant is absorbed onto the cellular structure—and an active mechanism—where a pollutant is bioaccumulated onto and within the living cell through respiration (Suresh Kumar et al., 2015). Selecting a microalgal species to achieve efficient bioremoval of pollutants requires consideration of variables such as strain, tolerance to the target pollutant, size, and life stage (Suresh Kumar et al., 2015). Although many species have been used for remediation over the years, the genera *Scenedesmus* and *Chlorella* appear to be the most extensively researched worldwide. *Scenedesmus* and *Chlorella* are the two most studied genera for attached microalgal cultivation, which facilitates harvesting and enhances water treatment efficiency (Zhuang et al., 2018). In addition to the above-mentioned genera, Table 4.1 summarizes other genera of interest, wastewater sources, and pollutant targets studied so far.

Scenedesmus obliquus is widely researched for WWT due to its mixotrophic characteristics. However, its application in real wastewater has been reported as unfavorable (Shen et al., 2015), since competition with ubiquitous bacteria may occur (Wang et al., 2013). Although this is a potential limitation, research conducted by Lin et al. (2017) demonstrated the species' efficiency in removing acetate (95.2%), propionate (97.1%), and butyrate (98.2%) present in starch-rich textile effluent. Similarly, Muñoz et al. (2019) assessed the application of *Scenedesmus* sp. to posttreat anaerobically digested food waste for subsequent use in irrigation. Their system reduced

TABLE 4.1 Microalgae Used for Different Wastewater Treatments

Microalgae	Wastewater Source	Pollutant Target	References
Chlamydomonas sp. TRC-1	Textile dyeing mill	Color	Behl et al. (2020)
Coelastrella sp. FI69	Liquefied petroleum gas–combusted exhaust gas streams	CO_2	Suriya Narayanan et al. (2019)
Desmodesmus armatus	Radionuclide-containing wastewater	Cesium	Kim et al. (2020)
Phaeodactylum tricornutum	Oxytetracycline-contaminated seawater	Oxytetracycline	Santaeufemia et al. (2016)
Picochlorum oculatum	Landfill leachate	NO_3–N and PO_4–P	Dogaris et al. (2019)
Tetraselmis indica	Anaerobically pretreated pharmaceutical wastewater	Chemical oxygen demand (COD), total organic carbon (TOC), nitrate, phosphate, and heavy metals	Amit et al. (2020)

COD by up to 97.5%. In addition, Yang et al. (2018) developed an algal biofilm microbial fuel cell for nutrient removal from domestic wastewater using *Scenedesmus quadricauda*. They achieved removal efficiencies of 96.0%, 91.5%, and 80.2% for TN, TP, and COD, respectively, and an algal density exceeding 3×10^4 cells mL^{-1} after 64 days of operation.

Wang et al. (2019) evaluated a three-stage process to treat anaerobically digested swine manure. The process comprised (i) flocculation and struvite precipitation to reduce P content, followed by (ii) an algal-bacterial (*Chlorella vulgaris* + sewage sludge) reactor; and (iii) activated carbon precipitation. The treatment unit achieved COD, TN, NH_4^+–N, and TP removal efficiencies of 97.2%, 94%, 99.9%, and 99.7%, respectively. Moreover, combining carbon-rich raw wastewater with nutrient-rich anaerobically digested wastewater from the food processing industry has been reported as a suitable solution for *in situ* WWT (Gupta and Pawar, 2018).

Cultivation of *Scenedesmus* sp. in raw wastewater from a WWT plant resulted in complete NH_3–N removal, along with 93% removal of TN and 61% removal of orthophosphate (Oliveira et al., 2018). Similar results were observed by Gupta, Choi et al. (2016) in a raw sewage treatment unit, where a mixotrophic cultivation system (*Chlorella vulgaris* and *Nannochloropsis oculata*) removed 80.62% of TN and 60.72% of total phosphate. In another study, cultivation of *Scenedesmus* sp. in tannery wastewater showed a maximum biomass concentration of 0.90 g L^{-1} and maximum removal efficiencies of ammonia nitrogen, phosphorus, and COD of 86%, 97%, and 80%, respectively (da Fontoura et al., 2017).

The use of stepwise cultivation also appears to benefit microalgal growth in wastewater. In this approach, the choice of microalgae must be based on the characteristics of each microalga; that is, one microalga should enhance the growth conditions for the next. Han et al. (2018) showed that the genera *Micractinium* and *Scenedesmus* reduce dissolved organic carbon in wastewater when cultivated in a two-step treatment system. The genus with the highest tolerance was cultivated in highly stressful wastewater to convert excess nutrients and reduce detrimental metal ions, thereby creating a suitable environment for the subsequent, less tolerant microalga.

Some microalgae can even treat waters contaminated with endocrine-disrupting chemicals (EDCs). For instance, Wang et al. (2019) investigated the removal of three estrogens—estrone, 17β-estradiol, and 17α-ethynylestradiol—by *Haematococcus pluvialis*, *Selenastrum capricornutum*, *Scenedesmus quadricauda*, and *Chlorella vulgaris*. They proposed six different degradation pathways for these estrogens, resulting in 11 distinct ring-cleavage products. Thus, it appears that even common microalgae can perform biotransformation and reduce the estrogenic activity of EDCs in wastewater.

Additionally, Luo et al. (2020) demonstrated the capability of the microalgae *Pseudokirchneriella subcapitata*, *Chlorella vulgaris*, and *Scenedesmus obliquus* to degrade six types of methylphenanthrenes (1-methylphenanthrene, 2-methylphenanthrene, 3-methylphenanthrene, 4-methylphenanthrene, 9-methylphenanthrene, and 3,6-dimethylphenanthrene). Remarkably, *P. subcapitata* removed 99.8% of 1-methylphenanthrene (1-MP) and 75.6% of 3,6-dimethylphenanthrene (3,6-DMP) after 7 days of incubation.

As mentioned above, *Chlorella* is adaptable to highly contaminated environments; therefore, its reliability is expected in different wastewater bioremediation applications. Indeed, *Chlorella* and *Scenedesmus* are often predominant in wastewater, as reported by Canovas et al. (1996). *Chlorella* is tolerant of various environmental conditions and is a strong candidate for remediation combined with by-product generation, such as biohydrogen (Jiménez-Llanos et al., 2020). Xie et al. (2018) evaluated the treatment of anaerobically digested effluent using an algal consortium. The predominant genus in the PBR was *Chlorella*, which was capable of removing nitrogen and phosphorus. In another study, Riaño et al. (2016) evaluated the potential of *Chlorella sorokiniana* to treat anaerobically digested cheese whey, achieving 84%–94% removal of soluble organic matter and ammonium. *Chlorella* has been described by Tan et al. (2019) as a suitable biotransformation vector for organic carbon. They cultivated *Chlorella pyrenoidosa* in starch-rich anaerobically digested effluent, reducing organic matter while increasing lipid production. Likewise, Nanda et al. (2019) demonstrated that *Chlorella sorokiniana* could grow even in wastewater containing organophosphate pesticides, and its lipid production was enhanced under exposure to malathion. Hence, the use of different species may represent a feasible approach for the remediation of contaminated water. However, it is important to carefully select the appropriate species based on the characteristics of the wastewater and the target compound.

4.3 MICROALGAE REMEDIATION SYSTEMS

Recently, many researchers have presented an interest in applying microalgae with cultivation in different wastewater sources (Gupta and Pawar, 2018; Hemalatha et al., 2019; Oliveira et al., 2018; Wang et al., 2019). Sometimes the studies have also investigated applications of the biomass generated, seeking to solve environmental problems and achieve economic yield through these techniques.

Studies performed by Hemalatha et al. (2019) indicated the potential use of microalgae in treating dairy wastewater and in producing multiple biobased products in a biorefinery framework, aiding in closed-loop recycling and avoiding the generation of waste streams. Nagarajan et al. (2019) investigated the application of microalgal cultivation for phytoremediation of swine wastewater due to its tolerance toward high ammonia levels and simultaneous phosphorus removal capacity. They highlighted the benefits of integrating microalgae cultivation after anaerobic digestion in a circular economy concept, given that it can improve the economic feasibility of on-farm biogas plants. The process is more economically viable with efficient water reuse and energy recovery from biomass. Concerning the water resources crisis and the worsening of environmental pollution, the pressure for new WWT techniques that are both effective and provide financial returns increases. The use of cultivation of microalgae utilizing wastewater, which, as previously presented, proves to be sustainable with the double benefit of biomass generation and nutrient remediation. The microalgae-based WWT process design and operation are primary factors for the successful application of microalgae to achieve high nutrient recovery, biomass productivity, and cost-effective technologies (Li et al., 2019).

4.3.1 SYSTEM INTEGRATION

The integration of microalgae cultivation with other processes has attained popularity, and the association of microalgae with clean technologies has been increasingly studied. This is due to the possibility of promoting the removal of pollutants, obtaining bioproducts, and reducing environmental impacts regarding greenhouse gas emissions (de Souza et al., 2019). Integrating microalgae-based processes into conventional tertiary treatment systems for nutrients is a cost-effective alternative for removing nutrients, a common challenge of treatment systems. Moreover, microalgae cultivation increases oxygen concentration and pH, provides disinfectant conditions, and allows the use of the biomass generated for secondary product development such as biofuels, fertilizers, and biopolymers (Bellucci et al., 2020). Microalgae are an alternative to promote onsite treatment for rural sanitation when integrated with septic tanks. These systems are not efficient in removing eutrophic loads and pathogens and require post-treatment (de Souza et al., 2019). Some system integrations that guarantee better results for removing contaminants and nutrients can be seen in Figure 4.3.

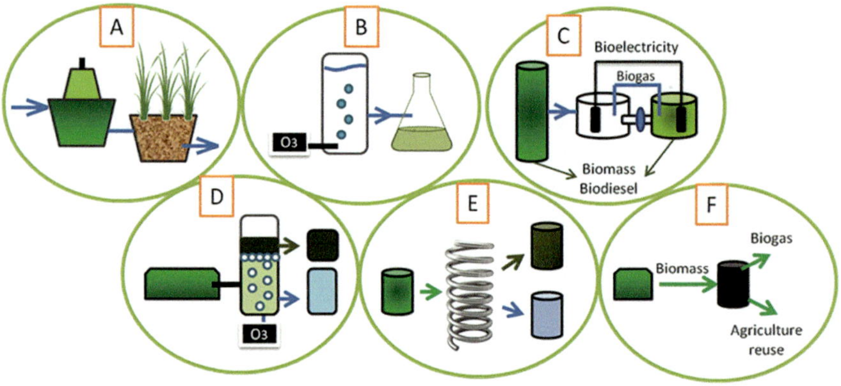

FIGURE 4.3 Different microalgae-integrated systems. (A) Microalgae tank and constructed wetland system; (B) ozonation followed by phycoremediation; (C) microalgae and microbial fuel cell for biomass and energy production; (D) algae harvesting through ozone flotation; (E) algae separation through electroflotation; and (F) application of digestate from microalgae anaerobic digestion with sewage sludge for agricultural reuse.

Source: Figures adapted from (A) Silveira et al. (2020); (B) Saranya and Shanthakumar (2020); (C) Nayak and Ghosh (2019); (D) Oliveira et al. (2018); (E) Baierle et al. (2015); and (F) Solé-Bundó et al. (2017).

The harvesting technique is still considered one of the bottlenecks and main limiting factors in WWT through microalgae cultivation (Molinuevo-Salces et al., 2019). In this context, the microalgae separation from the water process should be selected according to the algae characteristics and the biomass application's purpose. If the objective is treating wastewater, for example, the addition of a settler unit is sufficient. However, if biomass recovery is aimed, other additional systems should be applied, such as flotation, electrocoagulation, centrifugation, sedimentation, and filtering (Gupta et al., 2016), or the combination of different processes (Molinuevo-Salces et al., 2019). Baierle et al. (2015), for example, developed an innovative biomass separation system through electro flotation with a spiral-shaped electrode and using aluminum or iron tubes to identify the metal contamination present in the biomass and effluent. The authors obtained the best results with the aluminum electrodes; however, further investigations were recommended using this technique for microalgae separation from water.

Research on the use of microalgae for water quality improvement began in the 1950s (de Souza et al., 2019), and new combinations have been developed to increase biomass productivity, enhance biomass harvesting, promote bioelectricity production (microbial fuel cell systems), reduce treatment area requirements, and improve total water disinfection, among other applications. Table 4.2 presents a summary of current microalgae cultivation technologies and processes.

4.3.2 MICROALGAE-BACTERIA CONSORTIUM FOR WASTEWATER BIOREMEDIATION

Biological-based WWT by microorganisms (bacteria, algae, and fungi) can simulate natural ecosystems' ability to attenuate water pollution cost-effectively and sustainably (Ahmed et al., 2017). A consortium of microalgae with cyanobacteria and bacteria is more efficient for detoxifying organic and inorganic pollutants and removing nutrients from wastewater than individual microorganisms (Xiong et al., 2018). Furthermore, it is attractive due to its shorter hydraulic retention time (HRT), high cost-effectiveness, low energy consumption, and good settleability (Sun et al., 2019).

The use of microalgal-bacterial cultivation systems in wastewater bioremediation has been developed from the idea of exploiting the symbiotic

TABLE 4.2 Summary of Recent Studies Investigating Integrated Systems in the Literature

Country	Application	System Scale	Integrated Processes Description	References
Brazil	Urban wastewater treatment	Pilot	Microalgae combined with constructed wetlands (CW), aiming to reduce area demand, with CW acting as a filter for algal biomass.	Silveira et al. (2017)
Brazil	Urban wastewater treatment	Pilot	Biomonitoring of urban wastewater treated by an integrated system combining microalgae and constructed wetlands compared with a traditional wastewater treatment plant (WWTP).	Silveira et al. (2020)
India	Tannery effluent	Lab	Pretreatment with ozonation followed by phycoremediation to reduce costs and improve system feasibility.	Saranya and Shanthakumar (2020)
India	Pharmaceutical wastewater	Lab	Photobioreactor used as pretreatment integrated with a photosynthetic microbial fuel cell for the production of biomass and bioenergy (bioelectricity and biodiesel).	Nayak and Ghosh (2019)
Mexico	Urban wastewater treatment	Pilot	Ozone flotation applied in a high-rate algal pond for biomass recovery and improved nutrient removal.	Oliveira et al. (2018)
Italy	Urban wastewater treatment	Lab	Microalgae used as tertiary (polishing) treatment for disinfection and nutrient recovery from wastewater.	Bellucci et al. (2020)
Spain	Urban wastewater treatment	Pilot	Microalgal biomass from a treatment pond harvested from a secondary settler, preheated at 75°C to increase biodegradability, and then anaerobically digested and co-digested with sewage sludge for biogas production and agricultural reuse.	Solé-Bundó et al. (2017)

relationship between them. Application for wastewater bioremediation could be easily retrofitted into the existing advanced WWT facilities both at the secondary and tertiary levels of treatment (Gonçalves et al., 2016; Leong et al., 2019; Tang et al., 2016; Vu and Loh, 2016).

The relationship between the algae and bacteria strongly depends on the limiting factor for growth. Algae can produce oxygen and exudate bacteria's primary carbon sources (carbohydrates, proteins, organic acids, and lipids) (Norvill et al., 2016). In turn, bacteria can decompose organic matter into mineral form and secrete extracellular metabolites, such as auxins and vitamin B_{12}, besides CO_2, which are essential for microalgal growth. In addition, the cell surfaces of microalgae can provide a stable habitat for the bacteria (Xiong et al., 2018). The extracellular sheath (composed of various sugars) in some algae provides sites of attachment, organic carbon sources for bacterial growth, and photosynthetic oxygen as an electron acceptor for aerobic respiration. Similarly, some bacteria's siderophores can promote microalgae attachment and growth, especially in iron-deficient conditions (Wang et al., 2016).

Recent studies have demonstrated the use of algae-bacteria consortia for WWT, nutrient uptake, and organic matter removal. Foladori et al. (2018) achieved removal efficiencies of 87% ± 5% for COD and 98% ± 2% for TN without external aeration using photo-sequencing batch reactors (PSBRs) in municipal WWT. Similarly, García et al. (2017) demonstrated an innovative anoxic-aerobic algal-bacterial (*Scenedesmus*–Proteobacteria) PBR coupled with biogas upgrading for the treatment of domestic wastewater. Their system removed 81% of TN, 97% of NH_4^+, 64% of P–PO_4^{3-}, and 90% of TOC. Mohd. Udaiyappan et al. (2020) highlighted that the total amount of nutrients removed by the combination of microalgae and bacteria occurs through biological processes, such as assimilation and nitrification, as well as through physicochemical reactions, such as ammonia stripping and phosphate precipitation.

Microalgae-bacteria symbiosis is highly beneficial for organic carbon removal from high COD wastewater, since photosynthetic oxygenation provides a continuous O_2 supply for bacterial heterotrophic carbon metabolism (Nagarajan et al., 2020). Microalgae can also assist in co-metabolic degradation or produce biosurfactants and extracellular matters to enhance bacterial activity for increasing pollutant bioavailability (Tang et al., 2010). Bacteria might produce extracellular polymeric substances (EPS) that mediate their aggregation with microalgae and cyanobacteria, resulting in microalgal-bacterial aggregates (MABAs), which can increase the order of magnitude of the size of biomass particles three times. MABAs present sizes ranging from 100 to 5000 μm and dramatically improve absorption and sedimentation mechanisms in WWT approaches (Quijano et al., 2017).

The co-cultivation of algae and bacteria reduces high costs and decreases the spatial distance for O_2 and CO_2 exchange compared to separate culture units (Wang et al., 2016). Thus, the microalgae–bacterial sewage treatment process could be a significant alternative technology to aeration-based technologies like activated sludge treatment, thanks to the ability of the microalgae to re-aerate the medium (Ramanan et al., 2016).

Co-culture or consortia can perform functions that are limited or even impossible for individual strains or species. Living together may provide robustness to environmental fluctuations, stability for members, the ability to share metabolites during periods of nutrient limitation, resistance to invasion by other species, improved growth, and a greater potential for removing chemical contaminants from wastewater (Ma et al., 2014; Subashchandrabose et al., 2011). Ji et al. (2018) observed that chlorophyll-*a* concentrations in algae-bacteria symbiotic systems consisting of *Chlorella vulgaris–Bacillus licheniformis* were higher than those in single-microalgae systems, both cultivated in synthetic wastewater, indicating improved microalgal growth and enhanced nutrient utilization.

Luo et al. (2019) improved WWT through bioaugmentation with *Bacillus* and Paenibacillaceae and reported a 1.2-fold increase in biomass production of *Desmodesmus* sp. Therefore, the addition of bacteria promoted higher microalgal growth in cocultivation than in *Desmodesmus* sp. CHX1 monoculture. In another study (Qi et al., 2018), cocultures of three bacterial species (*Exiguobacterium aurantiacum*, *Stenotrophomonas acidaminiphila*, and *Chryseobacterium scophthalmus*) and the microalga *Chlorella sorokiniana* L3 resulted in better performance than pure cultures of microalgae or bacteria for the removal of four organic contaminants (ethanol, butanol, acetic acid, and butyric acid). The removal efficiency of volatile fatty acids and alcohols in the best coculture case increased by 22.70% compared with pure cultures. Systematic studies on the interaction mechanisms of microalgae-bacteria in WWT facilitate the formation and maintenance of sustainable ecological symbiotic systems (Li et al., 2020).

Besides substrate exchange, cell-to-cell communication and horizontal gene transfer have become mechanisms of interest. According to Zhang et al. (2018), quorum sensing is a cell-to-cell communication mechanism involving the exchange of small signaling molecules to coordinate gene expression, regulate physiological behavior, and support various ecological functions. Similarly, bacteria and microalgae can transfer genes and incorporate them into their genomes, thereby acquiring adaptive advantages under specific environmental stresses (Allen et al., 2011). Wang et al. (2016) suggested that microalgae-bacteria consortia are dynamic and are usually divided into four stages:

1. **Stage 1:** Adsorption of microalgae onto the surface of flocculated sludge.
2. **Stage 2:** Attachment of nascent bacteria onto the surface of microalgae, mainly within the phycosphere on the algal cell surface and its immediate surroundings.
3. **Stage 3:** Uniform distribution of microalgae and bacteria within the pellet, accompanied by the growth of both biotic constituents.
4. **Stage 4:** Formation of a mature consortium (Wang et al., 2016), as shown in Figure 4.4.

Several successful studies addressing microalgae-bacteria-based systems have demonstrated the feasibility of using consortia instead of monocultures. These systems can remove a variety of pollutants beyond nutrients and organic compounds (Table 4.3) and potentially treat several types of wastewater, such as landfill leachate (Tighiri and Erkurt, 2019), oil-containing effluents (Mahdavi et al., 2015b), agro-industrial wastewater (Hernández et al., 2013), piggery wastewater (González et al., 2008), industrial wastewater (Safonova et al., 2004), and pulp and paper industry wastewater (Yel et al., 2002), among others.

Microalgal sensitivity to toxic pollutants makes them suitable as test organisms for assessing acute toxicity. However, their sensitivity tends to decline when associated with degrading bacteria that exhibit greater resistance to the substrate (Muñoz and Guieysse, 2006). For instance, axenic *Scenedesmus obliquus* exhibited no oil-degrading ability. However, *Scenedesmus obliquus* combined with a consortium of oil component-degrading bacteria, including *Sphingomonas*, *Burkholderia cepacia*, *Pseudomonas*, and *Pandoraea pnomenusa*, significantly enhanced degradation capacity, particularly for recalcitrant polycyclic aromatic hydrocarbons (Tang et al., 2010). According to Mahdavi et al. (2015a), the algae-bacteria consortium enhanced the detoxification process through

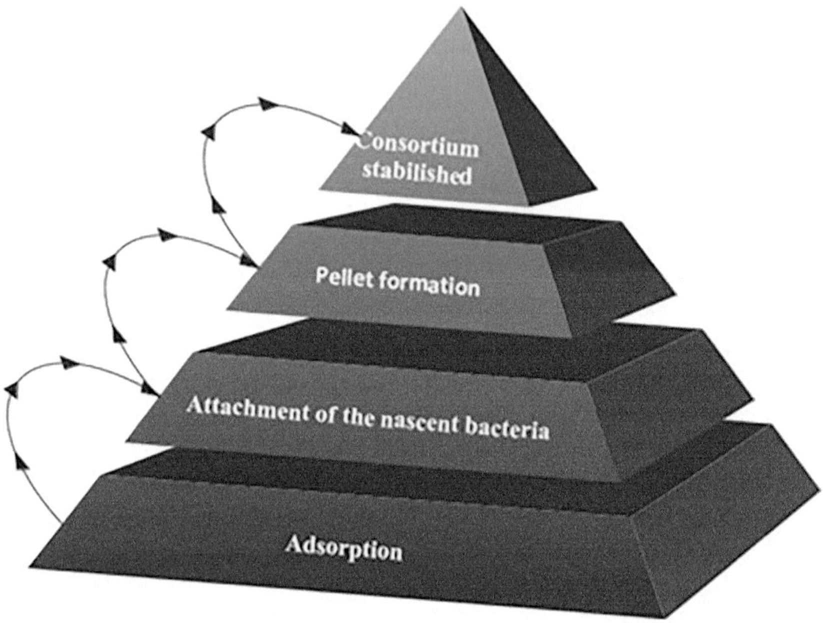

FIGURE 4.4 Representative illustration of the consortium stages.

Source: Adapted from Wang et al. (2016).

the bacterial biomass capacity to utilize functional groups and bonds (hydroxyl, carboxyl, and amide groups, as well as C–H, aryl–H, aryl–OH, and N–H bonds), thereby serving as potential markers for the biotransformation of toxic compounds.

The microalgal-bacterial consortia respond to toxic pollutants through various means, from physical adsorption, covalent bonding, biosorption, and bioaccumulation to biodegradation. The removal mechanism of hazardous compounds depends on many factors, such as the highlighted bioconcentration, degradation, temperature, pH, light supply, concentration, nutrient availability, dissolved oxygen concentration, biomass density, proportion of each microorganism, phylogenetic differences, and physiological function of the microorganism's communities (Mantovani et al., 2020; Zhang et al., 2020). Cometabolism assumes great importance in the biodegradation of bio-refractory pollutants in WWT approaches. The toxicity challenge of using biological processes can be overcome by adding a cometabolic substrate, such as acetate or formate, which can induce specific enzymes like monooxygenase and N-deethylase, involved

TABLE 4.3 Examples of Pollutant Removal Efficiency from Diverse Wastewater Types by Microalgae-Bacteria Consortia

Pollutant(s) Removed	Efficiency (%)	Wastewater Type	Bioreactor Type	Microalgae/Bacteria Consortium	References
Naphthalene, fluorene, and phenanthrene	100	Artificial medium containing petroleum-derived oil	Lab, 100 mL flasks	*Scenedesmus obliquus*/*Sphingomonas*, *Burkholderia cepacia*, *Pseudomonas*, and *Pandoraea pnomenusa*	Tang et al. (2010)
Lindane (γ-hexachlorocyclohexane (HCH))	100	Artificial medium containing lindane	Lab or pilot	*Scenedesmus* sp. ISTGA1 + *Paenibacillus* sp. ISTP10	Kumari et al. (2020)
Phenol	76.63–99.55	Landfill leachate	Photobioreactor	*Microcystis* sp., *Oscillatoria* sp., *Chlorella* sp., *Scenedesmus* sp., and *Stigeoclonium* sp./bacterial community	Tighiri and Erkurt (2019)
Salicylate	100	Artificial medium + salicylate	Lab, 155 mL flasks	*Chlorella sorokiniana*/*Ralstonia basilensis*	Borde et al. (2003)
Phenol	89			*Chlorella sorokiniana*/*Acinetobacter haemolyticus*	
Phenanthrene	23			*Chlorella sorokiniana*/*Pseudomonas migulae* and *Sphingomonas yanoikuyae*	
17β-estradiol	>93.75	Urban wastewater	Pilot-plant (indoor and outdoor) photobioreactor	*Scenedesmus* and *Chlorella*/bioaugmentation with bacterial community (Chlorophyta, Rotifera, Blastocladiomycota, Ciliophora, and Cercozoa)	Parladé et al. (2018)
Acetonitrile	100	Artificial wastewater	Pilot column photobioreactor	*Chlorella sorokiniana* and *Comamonas*	Muñoz et al. (2005)
Zn, Cr, Fe, Ni, and Pb	92, 91.5, 83.6, 69.2, and 74.9, respectively	Landfill leachate	Lab shake flask	*Scenedesmus* sp. ISTGA1 + *Paenibacillus* sp. ISTP10	Kumari et al. (2016)

in the degradation of toxic compounds (Xiong et al., 2018). Muñoz and Guieysse (2006) point out that resistant strains can be obtained by genetic manipulation, cell acclimation to progressively higher pollutant concentrations, or isolation from contaminated sites where indigenous microorganisms have already been exposed to the target contaminants to achieve better performance in both microbial growth and WWT approaches.

4.4 FUTURE PERSPECTIVES AND CHALLENGES OF WASTEWATER TREATMENT AND BIOMASS VALORIZATION

Several studies have feasibly redefined the dual role of microalgae, which can be used simultaneously to remedy wastewater and to obtain bioactive compounds. The market demand for food, feed, energy, and compounds with high added value, coupled with population increase, has made microalgae an important alternative for the bioeconomy (Sui et al., 2020). Recently, the microalgae ability for carbon sequestration is considered one of the most sustainable approaches for mitigating anthropogenic CO_2 and treating wastewater, with the possibility of using residual biomass to produce high-value compounds (Yadav et al., 2020).

The exploitation of microalgae biomass obtained from the clean medium has a market size of approximately 5000 t/year of dry matter, generating a capital return of 1.25×10^9 USD per year (Coêlho et al., 2019). Bioactive compounds produced by microalgae can result directly from primary metabolism or can be synthesized from secondary metabolism. In most microalgae, bioactive compounds are accumulated in the biomass; however, in some cases, these metabolites can be excreted into the environment (Mobin et al., 2019). The main bioactive compounds produced by microalgae are vitamins, lipids, proteins, pigments, antioxidants, and carbohydrates (de Souza et al., 2019). It is also noteworthy that microalgae production viability and sustainability on an industrial scale depend mainly on the successful development of biorefinery processes (Bhattacharya and Goswami, 2020; Coêlho et al., 2019). This potential is important for residual biomass from WWT by microalgae. Among the main advantages of focusing on studies in the scope of microalgae biomass valorization, the beneficial properties of these microorganisms are the most important since they can be applied in several fields. The main properties obtained from microalgae biomass are the presence of carotenoids, phycocyanin, polyphenols, fatty acids, and polyunsaturated compounds (Dvir et al., 2020;

Katiyar and Arora, 2020) and anticancer, anti-infectious, antioxidant, immunostimulant, and anti-inflammatory activities (Coêlho et al., 2019). These properties are currently used mainly in industries for animal fodder production and aquaculture, biofertilizers, natural food colors, cosmetics, drugs, and nutraceuticals/food products (Levasseur et al., 2020; Mobin et al., 2019; Oostlander et al., 2020). Therefore, the type of wastewater used in microalgae growth could be controlled by avoiding microbial contamination and residual toxic compounds.

The worldwide production of microalgae is still considered a fragile process and requires more studies due to the challenges regarding the high cost of production (Sui et al., 2020). Although the process's profitability based on microalgae is still questionable in some studies, sustainable and energy-efficient production will be not only crucial for environmental and commercial interests but will also be essential for advertising and to ensure, as a consequence, the mobilization in this area (Coêlho et al., 2019). According to Tang et al. (2020), the beneficial properties obtained through microalgae biomass (and incorporated into industrial products) can replace conventional synthetic materials and save costs. However, many challenges need to be overcome to commercialize these bioproducts, especially in the bioenergy area. The prospects for optimizing biomass processing and microalgae biorefineries will make it possible to reach the highly desired compounds of higher value and make remarkable progress towards microalgae-based biotechnology. Moreover, bioremediation technologies using microalgae-bacteria are suitable for biomass exploitation. Bacteria are naturally present in wastewater; therefore, studies on competition and synergism between these microorganisms and microalgae must be conducted to enhance WWT and increase biomass use (Gonçalves et al., 2017).

The nutrient recovery for subsequent use in the soil enables a new production stage of vegetable inputs and can be one of the main ways for cleaner production cycles. The biorefinery approach is the prime requirement for the sustainable production of biofertilizers after wastewater bioremediation using algae and microalgae (Castro et al., 2020). Huo et al. (2020) studied the effect of microalgae on plant growth and nutrient removal for three vegetables (arugula, purple kohlrabi, and lettuce) using nitrate-rich synthetic wastewater in greenhouse hydroponics. As a result, hydroponics was optimized by utilizing the co-production of microalgae with high value-added products in wastewater, using an environmentally and economically sustainable production system.

The feasibility of future advances using microalgae can be an excellent strategy for developing new technologies and obtaining economically viable processes. Figure 4.5 shows the main challenges that need to be overcome to use biomass to produce compounds of commercial interest properly. In this context, studies should focus on using microalgae or bacteria consortia that present an adequate profile to obtain the desired products. Using wastewater as a culture medium is an adequate option to valorize residues and mitigate the costs associated with nutrient supplementation. Subsequently, biomass exploitation must be evaluated, considering the different sectors that could be inserted (animal feed, fertilizers, biofuels). When there is no evidence of toxic substances, biomass can be directly used; however, if there is the presence of unregulated contaminants, methods should be evaluated to ensure that they can be safely removed. After the sequence of these parameters, cost evaluation techniques must be studied in terms of investments and earnings. It is noteworthy that the use of clean technologies can assist in optimizing this process to verify

FIGURE 4.5 Challenges in wastewater treatment using microalgae or microalgae-bacteria consortia with biomass exploitation.

alternative incomes for residues valorization in producing compounds of commercial interest with environmentally friendly processes.

Bioremediation followed by bioproducts development with algal technology is at an early stage. However, this possibility could mitigate possible regional excesses and substitute conventional high-impact products, enhancing sustainability within a circular economy (Stiles et al., 2018).

4.5 CONCLUSION

The bioremediation of wastewater by microalgae is an evolving field of research that currently seeks to identify microorganism species capable of removing recalcitrant molecules more efficiently through sustainable methods. In addition, this approach enables the synthesis of bioproducts based on the fundamental principles of the circular economy. It is already recognized that microalgae, whether used alone or in consortia, possess phycoremediation abilities for organic and inorganic molecules that are more difficult to remove from water. Moreover, this capability is independent of the equipment configuration. The use of microalgal biomass to produce high-value-added compounds remains one of the major challenges, since the biomass may contain molecules that restrict its use for food purposes. Depending on the production medium and the molecules removed, several bioproducts may be generated, including animal feed, fertilizers, and biofuels.

KEYWORDS

- **biofuels**
- **bioremediation**
- **exopolysaccharides**
- **fertilizers**
- **microalgae**
- **pentose phosphate pathway**
- **photobioreactors**
- **phycoremediation**
- **wastewater treatment**

REFERENCES

Agregán, R., Munekata, P. E. S., Franco, D., Carballo, J., Barba, F. J., & Lorenzo, J. M. (2018). Antioxidant potential of extracts obtained from macro- (*Ascophyllum nodosum, Fucus vesiculosus* and *Bifurcaria bifurcata*) and micro-algae (*Chlorella vulgaris* and *Spirulina platensis*) assisted by ultrasound. *Medicines, 5*(2), 33.

Ahmed, M. B., Zhou, J. L., Ngo, H. H., Guo, W., Thomaidis, N. S., & Xu, J. (2017). Progress in the biological and chemical treatment technologies for emerging contaminant removal from wastewater: A critical review. *Journal of Hazardous Materials, 323*, 274–298.

Allen, A. E., Dupont, C. L., Oborník, M., Horák, A., Nunes-Nesi, A., McCrow, J. P., et al. (2011). Evolution and metabolic significance of the urea cycle in photosynthetic diatoms. *Nature, 473*(7346), 203–207.

Amit, Nayak, J. K., & Ghosh, U. K. (2020). Microalgal remediation of anaerobic pretreated pharmaceutical wastewater for sustainable biodiesel production and electricity generation. *Journal of Water Process Engineering, 35*, 101192.

Asnaghi, V., Pecorino, D., Ottaviani, E., Pedroncini, A., Bertolotto, R. M., & Chiantore, M. (2017). A novel application of an adaptable modeling approach to the management of toxic microalgal bloom events in coastal areas. *Harmful Algae, 63*, 184–192.

Ata, A., Nalcaci, O. O., & Ovez, B. (2012). Macroalgae *Gracilaria verrucosa* as a biosorbent: A study of sorption mechanisms. *Algal Research, 1*(2), 194–204.

Baierle, F., John, D. K., Souza, M. P., Bjerk, T. R., Moraes, M. S. A., Hoeltz, M., et al. (2015). Biomass from microalgae separation by electro flotation with iron and aluminum spiral electrodes. *Chemical Engineering Journal, 267*, 274–281.

Behl, K., SeshaCharan, P., Joshi, M., Sharma, M., Mathur, A., Kareya, M. S., et al. (2020). Multifaceted applications of isolated microalgae *Chlamydomonas sp.* TRC-1 in wastewater remediation, lipid production and bioelectricity generation. *Bioresource Technology, 304*, 122993.

Bellucci, M., Marazzi, F., Naddeo, L. S., Piergiacomo, F., Beneduce, L., Ficara, E., et al. (2020). Disinfection and nutrient removal in laboratory-scale photobioreactors for wastewater tertiary treatment. *Journal of Chemical Technology and Biotechnology, 95*(4), 959–966.

Benedetti, M., Vecchi, V., Barera, S., & Dall'Osto, L. (2018). Biomass from microalgae: The potential of domestication towards sustainable bio factories. *Microbial Cell Factories, 17*(1), 173.

Beni, A. A., & Esmaeili, A. (2020). Biosorption, an efficient method for removing heavy metals from industrial effluents: A review. *Environmental Technology and Innovation, 17*, 100503.

Bhattacharya, M., & Goswami, S. (2020). Microalgae–A green multi-product biorefinery for future industrial prospects. *Biocatalysis and Agricultural Biotechnology, 25*, 101580.

Bilal, M., Rasheed, T., Sosa-Hernández, E. J., Raza, A., Nabeel, F., & Iqbal, M. N. H. (2018). Biosorption: An interplay between marine algae and potentially toxic elements—A review. *Marine Drugs, 16*(2), 65.

Borde, X., Guieysse, B., Delgado, O., Muñoz, R., Hatti-Kaul, R., Nugier-Chauvin, C., et al. (2003). Synergistic relationships in algal–bacterial microcosms for the treatment of aromatic pollutants. *Bioresource Technology, 86*(3), 293–300.

Bremner, C., Cochrane, T. A., McGuigan, P., & Bello-Mendoza, R. (2020). Removal of dissolved heavy metals from stormwater by filtration with granular recycled glass and mussel shell with and without microalgae biofilm. *Environmental Technology and Innovation, 18*, 100662.

Canovas, S., Picot, B., Casellas, C., Zulkifi, H., Dubois, A., & Bontoux, J. (1996). Seasonal development of phytoplankton and zooplankton in a high-rate algal pond. *Water Science and Technology, 33*(7), 199.

Castro, J. D. S., Calijuri, M. L., Ferreira, J., Assemany, P. P., & Ribeiro, V. J. (2020). Microalgae based biofertilizer: A life cycle approach. *Science of the Total Environment, 724*, 138138.

Chen, H., & Wang, Q. (2020). Microalgae-based nitrogen bioremediation. *Algal Research, 46*, 101775.

Coêlho, D. d. F., Tundisi, L. L., Cerqueira, K. S., Rodrigues, J. R. d. S., Mazzola, P. G., Tambourgi, E. B., et al. (2019). Microalgae: Cultivation aspects and bioactive compounds. *Brazilian Archives of Biology and Technology, 62*.

de Souza Celente, G., Colares, G. S., da Silva Araújo, P., Machado, Ê. L., & Lobo, E. A. (2020). Acute ecotoxicity and genotoxicity assessment of two wastewater treatment units. *Environmental Science and Pollution Research, 27*, 10520–10527.

de Souza Celente, G., Colares, G. S., Machado, Ê. L., & Lobo, E. A. (2019). Algae turf scrubber and vertical constructed wetlands combined system for decentralized secondary wastewater treatment. *Environmental Science and Pollution Research, 26*(10), 9931–9937.

de Souza, M. P., Hoeltz, M., Brittes Benitez, L., Machado, Ê. L., & de Souza Schneider, R. C. (2019). Microalgae and clean technologies: A review. *CLEAN–Soil, Air, Water, 47*(11), 1800380.

de Souza, M. P., Hoeltz, M., Gressler, P. D., Benitez, L. B., & Schneider, R. C. S. (2019). Potential of micro-algal bioproducts: General perspectives and main challenges. *Waste and Biomass Valorization, 10*(8), 2139–2156.

Delalat, B., Sheppard, V. C., Rasi Ghaemi, S., Rao, S., Prestidge, C. A., McPhee, G., et al. (2015). Targeted drug delivery using genetically engineered diatom biosilica. *Nature Communications, 6*(1), 8791.

Deruyck, B. (2019). Modeling the impact of rotifer contamination on microalgal production in open ponds, photobioreactor and thin layer cultivation systems. *Algal Research, 38*, 101398.

Devadasu, E., Chinthapalli, D. K., Chouhan, N., Madireddi, S. K., Rasineni, G. K., Sripadi, P., et al. (2019). Changes in the photosynthetic apparatus and lipid droplet formation in *Chlamydomonas reinhardtii* under iron deficiency. *Photosynthesis Research, 139*(1–3), 253–266.

Dogaris, I., Loya, B., Cox, J., & Philippidis, G. (2019). Study of landfill leachate as a sustainable source of water and nutrients for algal biofuels and bioproducts using the microalga *Picochlorum oculatum* in a novel scalable bioreactor. *Bioresource Technology, 282*, 18–27.

Droop, M. (1974). Heterotrophy of carbon. In *Algal Physiology and Biochemistry* (pp. 530–559).

Dvir, I., Moppes, D. V., & Arad, S. (2020). Foodomics: To discover the health potential of microalgae. In *Reference Module in Food Science*. Elsevier.

Fazal, T., Mushtaq, A., Rehman, F., Khan, A. U., Rashid, N., Farooq, W., et al. (2018). Bioremediation of textile wastewater and successive biodiesel production using microalgae. *Renewable and Sustainable Energy Reviews, 82*, 3107–3126.

Foladori, P., Petrini, S., & Andreottola, G. (2018). Evolution of real municipal wastewater treatment in photobioreactors and microalgae–bacteria consortia using real-time parameters. *Chemical Engineering Journal, 345*, 507–516.

Fu, W., Nelson, D. R., Mystikou, A., Daakour, S., & Salehi-Ashtiani, K. (2019). Advances in microalgal research and engineering development. *Current Opinion in Biotechnology, 59*, 157–164.

Furbank, R. T., & Tester, M. (2011). Phenomics–Technologies to relieve the phenotyping bottleneck. *Trends in Plant Science, 16*(12), 635–644.

García, D., Alcántara, C., Blanco, S., Pérez, R., Bolado, S., & Muñoz, R. (2017). Enhanced carbon, nitrogen and phosphorus removal from domestic wastewater in a novel anoxic–aerobic photobioreactor coupled with biogas upgrading. *Chemical Engineering Journal, 313*, 424–434.

Gonçalves, A. L., Pires, J. C. M., & Simões, M. (2016). Wastewater polishing by consortia of *Chlorella vulgaris* and activated sludge native bacteria. *Journal of Cleaner Production, 133*, 348–357.

Gonçalves, A. L., Pires, J. C. M., & Simões, M. (2017). A review on the use of microalgal consortia for wastewater treatment. *Algal Research, 24*, 403–415.

González, C., Marciniak, J., Villaverde, S., García-Encina, P. A., & Muñoz, R. (2008). Microalgae-based processes for the biodegradation of pretreated piggery wastewaters. *Applied Microbiology and Biotechnology, 80*(5), 891–898.

Gregory, J. A., Topol, A. B., Doerner, D. Z., & Mayfield, S. (2013). Alga-produced cholera toxin-Pfs25 fusion proteins as oral vaccines. *Applied and Environmental Microbiology, 79*(13), 3917–3925.

Gupta, P. L., Choi, H. J., & Lee, S. M. (2016). Enhanced nutrient removal from municipal wastewater assisted by mixotrophic microalgal cultivation using glycerol. *Environmental Science and Pollution Research, 23*(10), 10114–10123.

Gupta, P. L., Lee, S. M., & Choi, H. J. (2016). Integration of microalgal cultivation system for wastewater remediation and sustainable biomass production. *World Journal of Microbiology and Biotechnology, 32*(8), 139.

Gupta, S., & Pawar, S. B. (2018). An integrated approach for microalgae cultivation using raw and anaerobic digested wastewaters from food processing industry. *Bioresource Technology, 269*, 571–576.

Hammed, A. M., Prajapati, S. K., Simsek, S., & Simsek, H. (2016). Growth regime and environmental remediation of microalgae. *Algae, 31*(3), 189–204.

Han, J., Thomsen, L., Pan, K., & Thomsen, C. (2018). Two-step process: Enhanced strategy for wastewater treatment using microalgae. *Bioresource Technology, 268*, 608–615.

Hemalatha, M., Sravan, J. S., Min, B., & Venkata Mohan, S. (2019). Microalgae-biorefinery with cascading resource recovery design associated to dairy wastewater treatment. *Bioresource Technology, 284*, 424–429.

Hernández, D., Riaño, B., Coca, M., & García-González, M. C. (2013). Treatment of agro-industrial wastewater using microalgae–bacteria consortium combined with anaerobic digestion of the produced biomass. *Bioresource Technology, 135*, 598–603.

Hu, X., Meneses, Y. E., Stratton, J., & Wang, B. (2019). Acclimation of consortium of microalgae helps removal of organic pollutants from meat processing wastewater. *Journal of Cleaner Production, 214*, 95–102.

Huo, S., Liu, J., Addy, M., Chen, P., Necas, D., Cheng, P., et al. (2020). The influence of microalgae on vegetable production and nutrient removal in greenhouse hydroponics. *Journal of Cleaner Production, 243*, 118563.

Hussein, M. H., Abdullah, A. M., Badr El Din, N. I., & Mishaqa, E. S. I. (2017). Biosorption potential of the micro chlorophyte *Chlorella vulgaris* for some pesticides. *Journal of Fertilizers and Pesticides, 8*, 177.

Jasni, J., Narayanan Arisht, S., Mohd Yasin, N. H., Abdul, P. M., Lin, S. K., Liu, C. M., et al. (2020). Comparative toxicity effect of organic and inorganic substances in palm oil mill effluent (POME) using native microalgae species. *Journal of Water Process Engineering, 34*, 101165.

Ji, X., Jiang, M., Zhang, J., Jiang, X., & Zheng, Z. (2018). The interactions of algae–bacteria symbiotic system and its effects on nutrients removal from synthetic wastewater. *Bioresource Technology, 247*, 44–50.

Jiménez-Llanos, J., Ramírez-Carmona, M., Rendón-Castrillón, L., & Ocampo-López, C. (2020). Sustainable biohydrogen production by *Chlorella sp.* microalgae: A review. *International Journal of Hydrogen Energy, 45*(15), 8310–8328.

Katiyar, R., & Arora, A. (2020). Health promoting functional lipids from microalgae pool: A review. *Algal Research, 46*, 101800.

Kiki, C., Rashid, A., Wang, Y., Li, Y., Zeng, Q., Yu, C. P., et al. (2020). Dissipation of antibiotics by microalgae: Kinetics, identification of transformation products and pathways. *Journal of Hazardous Materials, 387*, 121985.

Kim, I., Choi, G. G., Nam, S. W., et al. (2020). Enhanced removal of cesium by potassium-starved microalga, *Desmodesmus armatus* SCK, under photoheterotrophic condition with magnetic separation. *Chemosphere, 252*, 126482.

Kumar, A., & Singh, J. S. (2017). Cyanoremediation: A green-clean tool for decontamination of synthetic pesticides from agro- and aquatic ecosystems. In *Agro-Environmental Sustainability* (pp. 59–83). Springer, Cham.

Kumari, M., Ghosh, P., & Thakur, I. S. (2016). Landfill leachate treatment using bacto-algal co-culture: An integrated approach using chemical analyses and toxicological assessment. *Ecotoxicology and Environmental Safety, 128*, 44–51.

Kumari, M., Ghosh, P., Swati, & Thakur, I. S. (2020). Development of artificial consortia of microalgae and bacteria for efficient biodegradation and detoxification of lindane. *Bioresource Technology Reports, 10*, 100415.

Leong, W. H., Zaine, S. N. A., Ho, Y. C., Uemura, Y., Lam, M. K., Khoo, K. S., et al. (2019). Impact of various microalgal-bacterial populations on municipal wastewater bioremediation and its energy feasibility for lipid-based biofuel production. *Journal of Environmental Management, 249*, 109384.

Leong, Y. K., & Chang, J. S. (2020). Bioremediation of heavy metals using microalgae: Recent advances and mechanisms. *Bioresource Technology, 303*, 122886.

Levasseur, W., Perré, P., & Pozzobon, V. (2020). A review of high value-added molecules production by microalgae in light of the classification. *Biotechnology Advances, 41*, 107545.

Li, K., Liu, Q., Fang, F., et al. (2019). Microalgae-based wastewater treatment for nutrients recovery: A review. *Bioresource Technology, 291*, 121934.

Lin, C. Y., Nguyen, M. L. T., & Lay, C. H. (2017). Starch-containing textile wastewater treatment for biogas and microalgae biomass production. *Journal of Cleaner Production, 168*, 331–337.

Litchman, E., Klausmeier, C. A., Schofield, O. M., & Falkowski, P. G. (2007). The role of functional traits and trade-offs in structuring phytoplankton communities: Scaling from cellular to ecosystem level. *Ecology Letters, 10*(12), 1170–1181.

Lu, W., Asraful Alam, M., Liu, S., Xu, J., & Parra Saldivar, R. (2020). Critical processes and variables in microalgae biomass production coupled with bioremediation of nutrients and CO_2 from livestock farms: A review. *Science of the Total Environment, 716*, 135247.

Luo, L. Z., Lin, X. A., Zeng, F. J., Wang, M., Luo, S., Peng, L., et al. (2019). Using co-occurrence network to explore the effects of bio-augmentation on the microalgae-based wastewater treatment process. *Biochemical Engineering Journal, 141*, 10–18.

Luo, L., Xiao, Z., Zhou, X., Yang, L., Luo, S., Zhao, C., et al. (2020). Quantum chemical calculation to elucidate the biodegradation pathway of methyl phenanthrene by green microalgae. *Water Research, 173*, 115598.

Ma, X., Zhou, W., Fu, Z., Cheng, Y., Min, M., Liu, Y., et al. (2014). Effect of wastewater-borne bacteria on algal growth and nutrients removal in wastewater-based algae cultivation system. *Bioresource Technology, 167*, 8–13.

Mahdavi, H., Prasad, V., Liu, Y., & Ulrich, A. C. (2015b). *In situ* biodegradation of naphthenic acids in oil sands tailings pond water using indigenous algae-bacteria consortium. *Bioresource Technology, 187*, 97–105.

Mantovani, M., Marazzi, F., Fornaroli, R., Bellucci, M., Ficara, E., & Mezzanotte, V. (2020). Outdoor pilot-scale raceway as a microalgae-bacteria side stream treatment in a WWTP. *Science of the Total Environment, 710*, 135583.

Mantzorou, A., Navakoudis, E., Paschalidis, K., & Ververidis, F. (2018). Microalgae: A potential tool for remediating aquatic environments from toxic metals. *International Journal of Environmental Science and Technology, 15*(8), 1815–1830.

Martínez-Bisbal, M. C., Carbó Mestre, N., Martínez-Máñez, R., Bauzá, J., & Alcañiz Fillol, M. (2019). Microalgae degradation follow up by voltametric electronic tongue, impedance spectroscopy and NMR spectroscopy. *Sensors and Actuators B: Chemical, 281*, 44–52.

Miranda, A. F., Ramkumar, N., Andriotis, C., Höltkemeier, T., Yasmin, A., Rochfort, S., et al. (2017). Applications of micro-algal biofilms for wastewater treatment and bioenergy production. *Biotechnology for Biofuels, 10*, 120.

Mobin, S. M. A., Chowdhury, H., & Alam, F. (2019). Commercially important bioproducts from microalgae and their current applications–A review. *Energy Procedia, 160*, 752–760.

Molinuevo-Salces, B., Riaño, B., Hernández, D., & Cruz García-González, M. (2019). Microalgae and wastewater treatment: Advantages and disadvantages. In *Microalgae Biotechnology for Development of Biofuel and Wastewater Treatment* (pp. 505–533). Springer Singapore.

Muñoz, P., Cordero, C., Tapia, X., Muñoz, L., & Candia, O. (2020). Assessment of anaerobic digestion of food waste at psychrophilic conditions and effluent post-treatment by microalgae cultivation. *Clean Technologies and Environmental Policy, 22*, 725–733.

Muñoz, R., & Guieysse, B. (2006). Algal–bacterial processes for the treatment of hazardous contaminants: A review. *Water Research, 40*(15), 2799–2815.

Muñoz, R., Rolvering, C., Guieysse, B., & Mattiasson, B. (2005). Photosynthetically oxygenated acetonitrile biodegradation by an algal-bacterial microcosm: A pilot-scale study. *Water Science and Technology, 51*(12), 261–265.

Murwanashyaka, T., Shen, L., Yang, Z., Chang, J. S., Manirafasha, E., Ndikubwimana, T., et al. (2020). Kinetic modeling of heterotrophic microalgae culture in wastewater: Storage molecule generation and pollutants mitigation. *Biochemical Engineering Journal, 157*, 107523.

Nagarajan, D., Kusmayadi, A., Yen, H. W., Dong, C. D., Lee, D. J., & Chang, J. S. (2019). Current advances in biological swine wastewater treatment using microalgae-based processes. *Bioresource Technology, 289*, 121718.

Nagarajan, D., Lee, D. J., Chen, C. Y., & Chang, J. S. (2020). Resource recovery from wastewaters using microalgae-based approaches: A circular bioeconomy perspective. *Bioresource Technology, 302*, 122817.

Nanda, M., Kumar, V., Fatima, N., Pruthi, V., Verma, M., Chauhan, P. K., et al. (2019). Detoxification mechanism of organophosphorus pesticide via carboxylesterase pathway that triggers de novo TAG biosynthesis in oleaginous microalgae. *Aquatic Toxicology, 209*, 49–55.

Nayak, J. K., & Ghosh, U. K. (2019). Post treatment of microalgae treated pharmaceutical wastewater in photosynthetic microbial fuel cell (PMFC) and biodiesel production. *Biomass and Bioenergy, 131*, 105415.

Neilson, A., & Lewin, R. (1974). The uptake and utilization of organic carbon by algae: An essay in comparative biochemistry. *Phycological, 13*(3), 227–264.

Nie, J., Sun, Y., Zhou, Y., Kumar, M., Usman, M., Li, J., et al. (2020). Bioremediation of water containing pesticides by microalgae: Mechanisms, methods, and prospects for future research. *Science of the Total Environment, 707*, 136080.

Nielsen, J., & Keasling, J. D. (2016). Engineering cellular metabolism. *Cell, 164*(6), 1185–1197.

Norvill, Z. N., Shilton, A., & Guieysse, B. (2016). Emerging contaminant degradation and removal in algal wastewater treatment ponds: Identifying the research gaps. *Journal of Hazardous Materials, 313*, 291–309.

Oliveira, G. A., Carissimi, E., Monje-Ramírez, I., Velasquez-Orta, S. B., Rodrigues, R. T., & Ledesma, M. T. O. (2018). Comparison between coagulation–flocculation and ozone–flotation for *Scenedesmus* microalgal biomolecule recovery and nutrient removal from wastewater in a high-rate algal pond. *Bioresource Technology, 259*, 334–342.

Oostlander, P. C., van Houcke, J., Wijffels, R. H., & Barbosa, M. J. (2020). Microalgae production cost in aquaculture hatcheries. *Aquaculture, 525*, 735310.

Pacheco, M. M., Hoeltz, M., Moraes, M. S. A., & Schneider, R. C. S. (2015). Microalgae: Cultivation techniques and wastewater phytoremediation. *Journal of Environmental Science and Health, Part A, 50*(6), 585–601.

Parladé, E., Hom-Diaz, A., Blánquez, P., Martínez-Alonso, M., Vicent, T., & Gaju, N. (2018). Effect of cultivation conditions on β-estradiol removal in laboratory and pilot-plant photobioreactors by an algal-bacterial consortium treating urban wastewater. *Water Research, 137*, 86–96.

Patel, A. K., Choi, Y. Y., & Sim, S. J. (2020). Emerging prospects of mixotrophic microalgae: Way forward to sustainable bioprocess for environmental remediation and cost-effective biofuels. *Bioresource Technology, 300*, 122741.

Perez-Garcia, O., Escalante, F. M., de-Bashan, L. E., & Bashan, Y. (2011). Heterotrophic cultures of microalgae: Metabolism and potential products. *Water Research, 45*(1), 11–36.

Qi, W., Mei, S., Yuan, Y., Li, X., Tang, T., Zhao, Q., et al. (2018). Enhancing fermentation wastewater treatment by co-culture of microalgae with volatile fatty acid- and alcohol-degrading bacteria. *Algal Research, 31*, 31–39.

Quijano, G., Arcila, J. S., & Buitrón, G. (2017). Microalgal-bacterial aggregates: Applications and perspectives for wastewater treatment. *Biotechnology Advances, 35*(6), 772–781.

Ramanan, R., Kim, B. H., Cho, D. H., Oh, H. M., & Kim, H. S. (2016). Algae–bacteria interactions: Evolution, ecology and emerging applications. *Biotechnology Advances, 34*(1), 14–29.

Rawat, I., Ranjith Kumar, R., Mutanda, T., & Bux, F. (2013). Biodiesel from microalgae: A critical evaluation from laboratory to large scale production. *Applied Energy, 103*, 444–467.

Riaño, B., Blanco, S., Becares, E., & García-González, M. C. (2016). Bioremediation and biomass harvesting of anaerobic digested cheese whey in microalgal-based systems for lipid production. *Ecological Engineering, 97*, 40–45.

Rodríguez-Vidal, F. J., García-Valverde, M., Ortega-Azabache, B., González-Martínez, Á., & Bellido-Fernández, A. (2020). Characterization of urban and industrial wastewaters using excitation–emission matrix (EEM) fluorescence: Searching for specific fingerprints. *Journal of Environmental Management, 263*, 110396.

Safonova, E., Kvitko, K. V., Iankevitch, M. I., Surgko, L. F., Afti, I. A., & Reisser, W. (2004). Biotreatment of industrial wastewater by selected algal-bacterial consortia. *Engineering in Life Sciences, 4*(4), 347–353.

Salama, E. S., Kurade, M. B., Abou-Shanab, R. A. I., El-Dalatony, M. M., Yang, I. S., Min, B., et al. (2017). Recent progress in microalgal biomass production coupled with wastewater treatment for biofuel generation. *Renewable and Sustainable Energy Reviews, 79*, 1189–1211.

Saranya, D., & Shanthakumar, S. (2020). An integrated approach for tannery effluent treatment with ozonation and phycoremediation: A feasibility study. *Environmental Research, 183*, 109163.

Serejo, M. L., Franco Morgado, M., García, D., González-Sánchez, A., Méndez-Acosta, H. O., & Toledo-Cervantes, A. (2020). Environmental resilience by microalgae. In *Microalgae Cultivation for Biofuels Production* (pp. 293–315). Academic Press.

Serrà, A., Artal, R., García-Amorós, J., Gómez, E., & Philippe, L. (2020). Circular zero-residue process using microalgae for efficient water decontamination, biofuel production, and carbon dioxide fixation. *Chemical Engineering Journal, 388*, 124278.

Shen, L., Wang, J., Li, Z., Fan, L., Chen, R., Wu, X., et al. (2020). A high-efficiency Fe_2O_3@microalgae composite for heavy metal removal from aqueous solution. *Journal of Water Process Engineering, 33*, 101026.

Shen, Q. H., Jiang, J. W., Chen, L. P., Cheng, L. H., Xu, X. H., & Chen, H. L. (2015). Effect of carbon source on biomass growth and nutrients removal of *Scenedesmus obliquus* for wastewater advanced treatment and lipid production. *Bioresource Technology, 190*, 257–263.

Silveira, E. O., Lutterbeck, C. A., Machado, Ê. L., Ribeiro Rodrigues, L., Rieger, A., Beckenkamp, F., et al. (2020). Biomonitoring of urban wastewaters treated by an integrated system combining microalgae and constructed wetlands. *Science of the Total Environment, 705*, 135864.

Silveira, E., Moura, D., Rieger, A., Machado, Ê., & Lutterbeck, C. (2017). Performance of an integrated system combining microalgae and vertical flow constructed wetlands for urban wastewater treatment. *Environmental Science and Pollution Research, 24*, 20469–20478.

Solé-Bundó, M., Cucina, M., Folch, M., Tàpias, J., Gigliotti, G., Garfí, M., et al. (2017). Assessing the agricultural reuse of the digestate from microalgae anaerobic digestion and co-digestion with sewage sludge. *Science of the Total Environment, 586*, 1–9.

Stiles, W. A. V., Styles, D., Chapman, S. P., Esteves, S., Bywater, A., Melville, L., et al. (2018). Using microalgae in the circular economy to valorize anaerobic digestate: Challenges and opportunities. *Bioresource Technology, 267*, 732–742.

Subashchandrabose, S. R., Ramakrishnan, B., Megharaj, M., Venkateswarlu, K., & Naidu, R. (2011). Consortia of cyanobacteria/microalgae and bacteria: Biotechnological potential. *Biotechnology Advances, 29*(6), 896–907.

Sui, Y., Jiang, Y., Moretti, M., & Vlaeminck, S. E. (2020). Harvesting time and biomass composition affect the economics of microalgae production. *Journal of Cleaner Production, 259*, 120782.

Sun, J., & Simsek, H. (2017). Bioavailability of wastewater derived dissolved organic nitrogen to green microalgae *Selenastrum capricornutum, Chlamydomonas reinhardtii,* and *Chlorella vulgaris* with/without presence of bacteria. *Journal of Environmental Sciences, 57*, 346–355.

Sun, L., Zuo, W., Tian, Y., Zhang, J., Liu, J., Sun, N., et al. (2019). Performance and microbial community analysis of an algal-activated sludge symbiotic system: Effect of activated sludge concentration. *Journal of Environmental Sciences, 76*, 121–132.

Sun, X. M., Ren, L. J., Zhao, Q. Y., Ji, X. J., & Huang, H. (2019). Enhancement of lipid accumulation in microalgae by metabolic engineering. *Biochemical et Biophysical Acta (BBA)–Molecular and Cell Biology of Lipids, 1864*(4), 552–566.

Suresh Kumar, K., Dahms, H. U., Won, E. J., Lee, J. S., & Shin, K. H. (2015). Microalgae–A promising tool for heavy metal remediation. *Ecotoxicology and Environmental Safety, 113*, 329–352.

Suriya Narayanan, G., Kumar, G., Seepana, S., Elankovan, R., & Premalatha, M. (2019). Utilization of unfiltered LPG-burner exhaust-gas emission using microalga *Coelastrella sp. Journal of* CO_2 *Utilization, 29*, 283–295.

Sutherland, D. L., & Ralph, P. J. (2019). Microalgal bioremediation of emerging contaminants—Opportunities and challenges. *Water Research, 164*, 114921.

Tan, X. B., Zhao, X. C., & Yang, L. B. (2019). Strategies for enhanced biomass and lipid production by *Chlorella pyrenoidosa* culture in starch processing wastewater. *Journal of Cleaner Production, 236*, 117671.

Tang, C. C., Zuo, W., Tian, Y., Sun, N., Wang, Z. W., & Zhang, J. (2016). Effect of aeration rate on performance and stability of algal-bacterial symbiosis system to treat domestic wastewater in sequencing batch reactors. *Bioresource Technology, 222*, 156–164.

Tang, D. Y. Y., Khoo, K. S., Chew, K. W., Tao, Y., Ho, S. H., & Show, P. L. (2020). Potential utilization of bioproducts from microalgae for the quality enhancement of natural products. *Bioresource Technology, 304*, 122997.

Tang, X., He, L. Y., Tao, X. Q., Dang, Z., Guo, C. L., Lu, G. N., et al. (2010). Construction of an artificial microalgal-bacterial consortium that efficiently degrades crude oil. *Journal of Hazardous Materials, 181*(1–3), 1158–1162.

Tanner, W. (2000). The *Chlorella* hexose/H^+ symporters. *International Review of Cytology, 200*, 101–141.

Taunt, H. N., Stoffels, L., & Purton, S. (2018). Green biologics: The algal chloroplast as a platform for making biopharmaceuticals. *Bioengineered, 9*(1), 48–54.

Tighiri, H. O., & Erkurt, E. A. (2019). Biotreatment of landfill leachate by microalgae-bacteria consortium in sequencing batch mode and product utilization. *Bioresource Technology, 286*, 121396.

Tripathi, S., Arora, N., Gupta, P., Pruthi, P. A., Poluri, K. M., & Pruthi, V. (2019). Microalgae: An emerging source for mitigation of heavy metals and their potential implications for biodiesel production. In *Advanced Biofuels* (pp. 97–128). Elsevier, Woodhead Publishing.

Udaiyappan, A. F. M., Abu Hasan, H., Takriff, M. S., Sheikh Abdullah, S. R., Maeda, T., Mustapha, N. A., et al. (2020). Microalgae-bacteria interaction in palm oil mill effluent treatment. *Journal of Water Process Engineering, 35*, 101203.

Urrutia, C., Yañez-Mansilla, E., & Jeison, D. (2019). Bio removal of heavy metals from metal mine tailings water using microalgae biomass. *Algal Research, 43*, 101659.

Vu, L. T. K., & Loh, K. C. (2016). Symbiotic hollow fiber membrane photobioreactor for microalgal growth and bacterial wastewater treatment. *Bioresource Technology, 219*, 261–269.

Wang, H., Zhang, W., Chen, L., Wang, J., & Liu, T. (2013). The contamination and control of biological pollutants in mass cultivation of microalgae. *Bioresource Technology, 128*, 745–750.

Wang, L., Addy, M., Liu, J., Nekich, C., Zhang, R., Peng, P., et al. (2019). Integrated process for anaerobically digested swine manure treatment. *Bioresource Technology, 273*, 506–514.

Wang, Y., Ho, S. H., Cheng, C. L., Guo, W. Q., Nagarajan, D., Ren, N. Q., et al. (2016). Perspectives on the feasibility of using microalgae for industrial wastewater treatment. *Bioresource Technology, 222*, 485–497.

Wang, Y., Sun, Q., Li, Y., Wang, H., Wu, K., & Yu, C. P. (2019). Biotransformation of estrone, 17β-estradiol and 17α-ethynyl estradiol by four species of microalgae. *Ecotoxicology and Environmental Safety, 180*, 723–732.

Weis, L., de Cassia de Souza Schneider, R., Hoeltz, M., Rieger, A., Tostes, S., & Lobo, E. A. (2020). Potential for bifenthrin removal using microalgae from a natural source. *Water Science and Technology, 82(6), 1131–1141.*

Xie, B., Gong, W., Tian, Y., Qu, F., Luo, Y., Du, X., et al. (2018). Biodiesel production with the simultaneous removal of nitrogen, phosphorus and COD in microalgal-bacterial communities for the treatment of anaerobic digestion effluent in photobioreactors. *Chemical Engineering Journal, 350*, 1092–1102.

Xiong, J. Q., Kurade, M. B., & Jeon, B. H. (2018). Can microalgae remove pharmaceutical contaminants from water? *Trends in Biotechnology, 36*(1), 30–44.

Xiu, B., Liang, S. K., He, X. L., Wang, X. K., Cui, Z. G., & Jiang, Z. J. (2019). Bioavailability of dissolved organic nitrogen and its uptake by *Ulva prolifera*: Implications in the outbreak of a green bloom off the coast of Qingdao, China. *Marine Pollution Bulletin, 140*, 563–572.

Yadav, G., & Sen, R. (2018). Sustainability of microalgal biorefinery: Scope, challenges, and opportunities. In *Sustainable Energy Technology and Policies: A Transformational Journey* (pp. 335–351). Springer Singapore.

Yadav, G., Meena, D. K., Sahoo, A. K., Das, B. K., & Sen, R. (2020). Effective valorization of microalgal biomass for the production of nutritional fish-feed supplements. *Journal of Cleaner Production, 243*, 118697.

Yang, Z., Pei, H., Hou, Q., Jiang, L., Zhang, L., & Nie, C. (2018). Algal biofilm-assisted microbial fuel cell to enhance domestic wastewater treatment: Nutrient and organics removal and bioenergy production. *Chemical Engineering Journal, 332*, 277–285.

Yel, E., Dilek, F., & Yetis, U. (2002). Effectiveness of algae in treatment of a wood-based pulp and paper industry wastewater. *Bioresource Technology, 84*, 1–5.

Zhan, J., Rong, J., & Wang, Q. (2017). Mixotrophic cultivation, a preferable microalgae cultivation mode for biomass/bioenergy production and bioremediation: Advances and prospect. *International Journal of Hydrogen Energy, 42*(12), 8505–8517.

Zhang, B., Lens, P. N. L., Shi, W., Zhang, R., Zhang, Z., Guo, Y., et al. (2018). The attachment potential and N-acyl-homoserine lactone-based quorum sensing in aerobic granular sludge and algal-bacterial granular sludge. *Applied Microbiology and Biotechnology, 102*(12), 5343–5353.

Zhang, B., Li, W., Guo, Y., Zhang, Z., Shi, W., Cui, F., et al. (2020). Microalgal-bacterial consortia: From interspecies interactions to biotechnological applications. *Renewable and Sustainable Energy Reviews, 118*, 109563.

Zhang, H., Gong, W., Zeng, W., Yan, Z., Jia, B., Li, G., et al. (2020). Organic carbon promotes algae proliferation in membrane-aeration based bacteria-algae symbiosis system (MA-BA). *Water Research, 176*, 115736.

Zhuang, L. L., Yu, D., Zhang, J., Liu, F. F., Wu, Y. H., Zhang, T. Y., et al. (2018). The characteristics and influencing factors of the attached microalgae cultivation: A review. *Renewable and Sustainable Energy Reviews, 94*, 1110–1119.

CHAPTER 5

Microalgae Cultivation in Algal Ponds for Wastewater Treatment

MRINAL KANTI ADAK[1,2,*] and AJIT DAS[1]

[1]*Department of Chemistry, Sidho-Kanho-Birsha University, Purulia, West Bengal, India*

[2]*Department of Chemistry, Indian Institute of Technology Delhi, New Delhi, Delhi, India*

*Corresponding author

ABSTRACT

Microalgae-based wastewater treatment (WWT) is one of the most promising processes for wastewater treatment, enabling high algal biomass production through a greener and more economical pathway. This approach is effective in terms of low cost and environmental sustainability for pollutant removal, along with reduced greenhouse gas emissions. Techniques for the removal of inorganic nutrients and hazardous organic substances are implemented through open ponds and enclosed photobioreactors (PBRs). This book chapter aims to provide a concise overview of microalgae cultivation in ponds for WWT and their performance under different environmental conditions. Previously reported studies discuss various types of microalgae cultivation in different pond systems for effective pollutant removal. The algal metabolic processes, including biodegradation and photodegradation in algal ponds for WWT, are also discussed. In the cultivation process, the degradation mechanism varies according to the nature of the wastewater and system parameters.

Environmental Phycology: Climate Change Mitigation and Hazardous Pollutant Cleanup Using Algal Resources. Jeyabalan Sangeetha and Devarajan Thangadurai (Eds.)

DOI: 10.1201/9781779647085-5

5.1 INTRODUCTION

In the modern era, wastewater treatment (WWT) is one of the major concerns globally for the living community. The major sources of wastewater arise from anthropogenic activities of humans, such as agricultural practices, urbanization, and industrialization (Aslan and Kapdan, 2006; Rawat et al., 2011; Renuka et al., 2013), causing widespread emerging contaminants. Untreated wastewater leads to an increase in organics, nutrients like nitrogen and phosphorus in water bodies (eutrophication phenomenon) (Gonçalves et al., 2017), changes medium pH, and decreases dissolved oxygen, which poses a serious risk to the aquatic ecosystem (Cai et al., 2019), including chronic ecotoxicity, antibiotic resistance, and the involvement of toxic pollutants in the food chain (Arnold et al., 2013; Ding and He, 2010).

Therefore, WWT is essential to sustain life, the environment, and ecosystems. The primary, secondary, tertiary, and quaternary stages of wastewater treatment are presented schematically in Figure 5.1. Conventional treatment processes, such as the activated sludge system, are widely used worldwide to treat wastewater; however, they contribute to the prevalence of antibiotic-resistant bacteria and antibiotic resistance genes (Cheng et al., 2020). Additionally, this system includes an aeration tank and a secondary settling tank and requires skilled operators, making it costly and complex, with high energy consumption (Berland and Cooper, 2001; Massoud et al., 2009). The subinhibitory effects of antibiotics play a vital role in the spread of antibiotic resistance by accelerating mutation rates and horizontal gene transfer (Baharoglu and Mazel, 2014). A significant correlation has been observed between the number of resistant bacteria and antibiotic concentration in wastewater (Gao et al., 2012). In activated sludge after treatment, the abundance of antibiotic resistance genes and antibiotic-resistant bacteria was reported as 1.00×10^8–1.78×10^9 gene copies/100 mL and 7.08×10^6–1.91×10^8 CFU/100 mL, respectively (Gao et al., 2012). This process also increases gene transfer via transduction due to the presence of more than 1000 unique viral genomes (Cheng et al., 2020).

In this respect, algal pond systems are more powerful in removing organics, heavy metals, coliform bacteria, pathogens, nutrients, and also biochemical oxygen demand (BOD). Algal-based technologies were introduced in the 1950s to solve the wastewater problem under a greener,

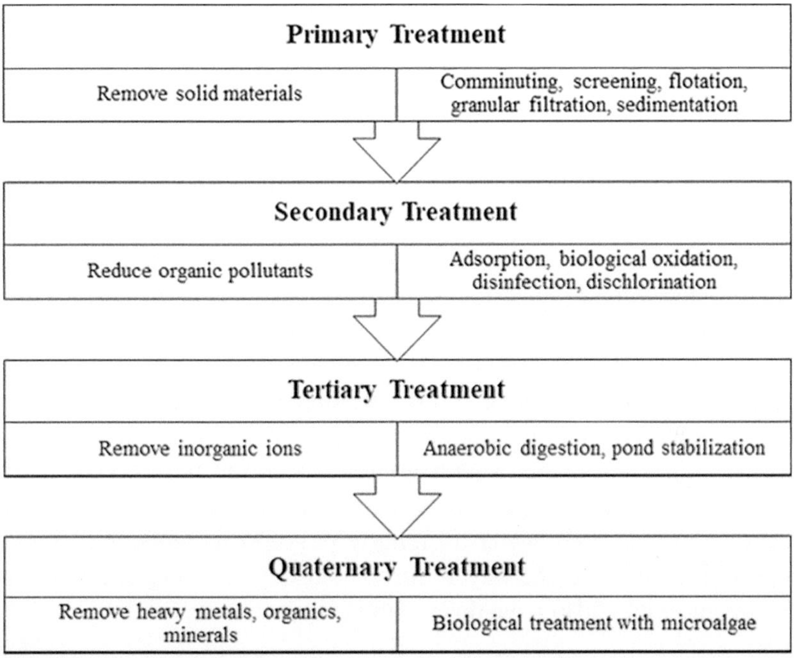

FIGURE 5.1 Schematic diagram of primary, secondary, tertiary, and quaternary treatment stages in a wastewater treatment plant.

long-term secure pathway (Delanka-Pedige et al., 2019). The main advantage of this technique is that the process does not require any additional mechanical aeration during WWT, which reduces its energy demand (Henkanatte-Gedera et al., 2015; Selvaratnam et al., 2014). Aeration is achieved through the photosynthetic O_2 release, which increases dissolved oxygen up to 400% saturation (Holmes et al., 2020). The technique also supports autotrophic and heterotrophic metabolisms, which enhance its activity. The enhancement of water alkalinity and the fixation of CO_2 by microalgae favor organic matter oxidation, making this technique simple, cost-effective, and sustainable for the environment (García et al., 2017). Microalgae effectively consume nitrogen and phosphorus for their growth (Munoz and Guieysse, 2006; Renuka et al., 2013), which makes it efficient for 80–100% removal of nutrients from wastewater (González et al., 1997; Li et al., 2011). Again, its high growth rate and efficiency rate are much higher by a factor of 10–50 than other terrestrial plants, and their growth in unfavorable environmental states, marginal land, or wasteland makes them unique (Hoh et al., 2016). The use of this combined tertiary and quaternary treatment method (Rayen et al., 2019; Salama et al., 2017) offers several advantages: (i) nutrients such as nitrogen and phosphorus, absorbed by microalgae, can be recycled through the production of fertilizer from the biomass; (ii) the biomass can be used for the production of biofuels, food, animal feed, pharmaceuticals, and bioplastics; (iii) the method discharges oxygenated effluent into receiving water bodies; and (iv) it provides a higher lipid content compared with feedstocks used for first- and second-generation biofuels (Gonçalves et al., 2017; Hoh et al., 2016). Moreover, this system can sequester biogenic CO_2 produced by bacteria and has great potential to reduce greenhouse gas emissions, making it a "carbon-negative" process (Holmes et al., 2020).

For a stable algal population in the WWT plant, the interaction between algae and bacteria with wastewater in the presence of light is very important. This chapter deals with the algal photosynthetic aeration technique to treat wastewater for long-term goals. As biodegradation and photodegradation are the key removal pathways for organic pollutants (Gentili and Fick, 2017), we discuss those in a separate section.

5.2 ALGAL PONDS FOR WASTEWATER TREATMENT

Algal ponds are stable, green, carbon negative, low energy input systems with a long hydraulic preservation time process to treat wastewater with high efficiency. Oswald and their group (Oswald, 1990) proposed in detail the design of an advanced integrated wastewater pond system (AIWPS), where lower energy input gives maximum removal efficiency. Rawat et al. (2016) reported that AIWPS consists of at least four basic ponds: a facultative pond with an internal fermentation pit, a high-rate algal pond, an algae-settling pond, and finally a maturation pond, each with different objectives. Tilley and coworkers described three types of ponds—(i) anaerobic; (ii) facultative; and (iii) aerobic (maturation)—in the 2nd revised edition of *Compendium of Sanitation Systems and Technologies* (Tilley et al., 2014). They presented these three pond types in the schematic diagram shown in Figure 5.2. According to Norvill et al. (2016), there are three main types of algal ponds: facultative ponds, maturation ponds, and high-rate algal ponds. The specifications of different algal ponds are given below:

1. **Facultative Ponds:** This is a shallow pond with a depth of 1 to 2.5 m and a detention time of 5 to 30 days. The surface layer consists

of an aerobic zone where oxygen (from algal photosynthesis, natural diffusion, and wind mixing) is consumed by aerobic bacteria, and a deeper anaerobic zone where digestion takes place (Tilley et al., 2014). Here, the overflow velocity is kept low (<1.5 m d^{-1}) to allow solids to settle (Rawat et al., 2016). The primary function is to stabilize organic matter, reduce nutrient load, and remove BOD from wastewater (Schwarz et al., 2019). A properly designed facultative pond with an internal fermentation pit can remove 100% of suspended solids and achieve 60–70% BOD removal (Rawat et al., 2016).

2. **Maturation Ponds:** This is an aerobic pond with a depth of 0.5 to 1.5 m and a retention time of 10–15 days (Rawat et al., 2016). These ponds are typically arranged in series to achieve higher pathogen removal. The addition of algae or fish harvesting can increase their efficiency. Here, sunlight should penetrate the full depth of the pond (Tilley et al., 2014). This type of pond is also effective for removing chromium, selenium, and other contaminants from wastewater (Rawat et al., 2016).

3. **High-Rate Algal Ponds:** "High-rate algal ponds" (HRAPs) were first introduced by Oswald and Golueke (Oswald and Golueke, 1960). They observed that mass culture of algae offers the possibility of fixing solar energy with a tenfold increase in areal yield in ponds. This natural technique is now successfully applied worldwide to treat wastewater, primarily for the removal of BOD and nutrients, and also for biofuel production (Craggs et al., 2013; Park and Craggs, 2010). The retention time for HRAPs is 4–10 days (Rawat et al., 2016). This type of pond is also known as a raceway pond. Simple operation and low cost make HRAPs suitable for rural, peri-urban, and remote communities (Acién et al., 2016; Young et al., 2017). In HRAPs, a paddlewheel is used to maintain horizontal flow velocities of 0.15–0.30 m s^{-1} (Chiaramonti et al., 2013) to enhance water mixing and photosynthetic capacity (Uggetti et al., 2018). HRAPs have an excellent capacity to treat wastewater through direct and indirect mechanisms to remove nutrients, phosphate, ammonia, and organic pollutants, which makes the system popular worldwide. This type of pond is also

useful for the treatment of effluents from tanneries, mines, zinc refineries, and piggery farms (Rawat et al., 2016).

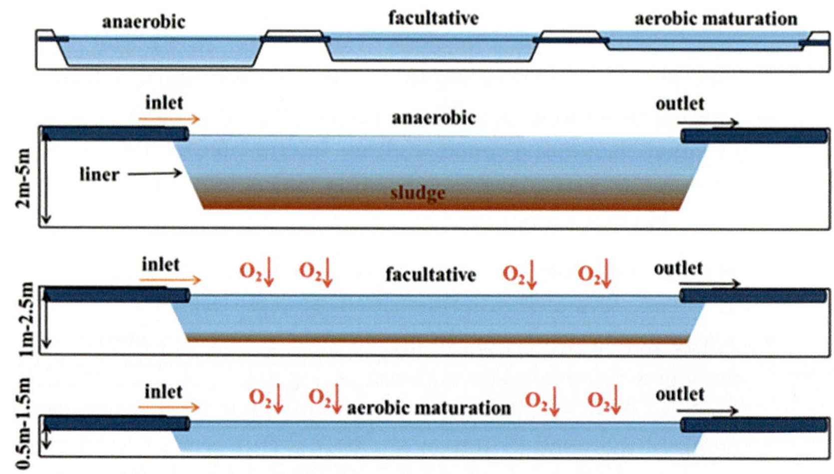

FIGURE 5.2 Design characteristics of waste stabilization ponds.

Again, there are two types of photo-bioreactors where microalgae can be cultivated. The first one is an open photo-bioreactor and the other is a close photo-bioreactor. Open bioreactors are generally HRAPs with a cost-effective solution and are used for large-scale applications. This type of bioreactors includes natural water sources shallow, big ponds, tanks, circular ponds, and raceway ponds (Suh and Lee, 2003). Close bioreactors are designed according to space requirements with better control of parameters like water loss, biomass, light, and so forth, which leads to cost expensive. In an open photo-bioreactor, there are several challenges like contamination of microalgae, low light intensity, day-night cycle, temperature variation, and also a large area of land (Yen et al., 2019). So, the main criteria for a microalgae cultivation system are light efficiency and mixing efficiency at a low cost. In an algal pond system, the surface area to volume ratio is very high with long hydraulic retention time (HRT) than in another conventional system. As a result, the system is more facilitated in light coverage and aerobic biodegradation by photosynthetic oxygen from algae. This dissolved oxygen (DO) is also used by aerobic bacteria to degrade organic pollutants. Aerobic bacteria have symbiotic relationships with microalgae capable of organic carbon removal (Garcia et al., 2000).

The algal pond system enhances DO in water, develops biomass settling capability, diminishes CO_2, and enlarges medium pH that enhances ammonia volatilization with phosphate precipitation (Wang et al., 2018). Microalgal metabolism also enhances bacterial growth due to the enhancement of growth-promoting factors. The oxygen, and organic acids are produced by algae for bacteria whereas bacteria also produce CO_2 and vitamins for algae, and they depend on each other known as a synergistic relationship (Munoz and Guieysse, 2006). In this respect, the ratio of algal and bacterial biomass is very important and is varying seasonally (Garcia et al., 2000; Oron et al., 1979). Temperature and light are two important parameters for a pond system, and they should be considered during the design of a pond. In hot geographical locations, the system suffers a large amount of water evaporation with CO_2 losses whereas in cold places algae growth is inhibited (Rayen et al., 2019). The utilization of an algal pond system that contains both WWT and biofuel production gives a bright future for an alternative green energy source.

5.3 CHARACTERIZATIONS OF WASTEWATER AND MICROALGAE

Wastewater can be defined as water containing dissolved or suspended wastes from various sources, including domestic, agricultural, urban, municipal, and industrial activities (Gonçalves et al., 2017). It contains several contaminants, such as macro- (C, N, P, H, O, S, Mg, K, Na, and Ca) and micro- (Fe, Cu, Mn, Zn, Cl, V, Mo, B, Co, and Si) nutrients, organic pollutants, pathogenic organisms, inorganic compounds, and sediments (Salama et al., 2017; Sonune and Ghate, 2004). The characteristics of wastewater vary according to its source. In the case of domestic wastewater, sewage contains approximately 99.9% water, whereas the remaining 0.1% consists of organic and inorganic substances, suspended and dissolved solids, and microorganisms (Sperling and Von, 2007). Agro-industrial wastewater is characterized by high ammonium concentrations and is strongly associated with eutrophication (Madakka et al., 2019). In municipal wastewater, nitrogen is present as ammonium, nitrate, nitrite, and organic nitrogen, while phosphorus occurs as orthophosphate and organic phosphorus, with variations in carbon-containing compounds (Kube et al., 2018). Effluents from industries such as carpet mills, dairy processing, piggery operations, paper manufacturing, chemical production, and breweries vary depending on the industry. The N:P molar ratio in effluents strongly influences microalgal biomass production. When the N:P ratio reaches 30:1, it indicates N limitation, whereas a low ratio of 5:1 leads to P limitation. For phytoplankton growth, the optimum N:P ratio is generally 15:1 (Darley, 1982).

Microalgal growth depends on the substantial concentration of nutrients, mainly carbon, nitrogen, and phosphorus, and in the presence of suitable pH and dissolved CO_2, which are related to the characteristics of wastewater (Abdel-Raouf et al., 2012). In this manner, a good understanding is required between microalgal growth and the wastewater medium. The growth limitation and stoichiometric ratio of algal biomass depend on the molar ratio of carbon, nitrogen, and phosphorus in wastewater (Salama et al., 2017). For example, the secondary effluent contains low-level organic substances that are useful for membrane photo-bioreactor feed. In this case, a high organic carbon level is harmful due to heterotrophic bacteria growth competing with microalgae (Pittman et al., 2011; Xu et al., 2015). The restriction also arises from high ammonia levels, which inhibit microalgal growth. A reported study showed that the growth of *Scenedesmus* is optimal up to an ammonia concentration of 100 mg/L; however, at 200–500 mg/L, the biomass is reduced by up to 70% (Park et al., 2010). *Nephroselmis pyriformis* has been shown to exhibit toxicity to unionized ammonia that is more than 100 times greater than that of its ionized form (Källqvist and Svenson, 2003). The degradation of organic pollutants from the oil and paper industries has been successfully achieved in the presence of microalgae such as *Chlorella*, *Ankistrodesmus*, and *Scenedesmus* species (Abeliovich and Weisman, 1978; Pinto et al., 2002, 2003; Tarlan et al., 2002).

5.3.1 ALGAL METABOLISM

Microalgae can be classified into three categories according to their metabolic pathways involved in pollutant removal during cultivation: (i) photoautotrophy; (ii) heterotrophy; and (iii) mixotrophy (Nirmalakhandan et al., 2019). Among these, photoautotrophy is more economical and environmentally friendly than the others. The functions and applications of each type are given below:

1. **Photo-Autotrophy:** In phototrophic or autotrophic culture light is used as an energy source and anabolic photosynthetic CO_2 as a carbon source for the growth of microalgae. This type of culture

has been used in two ways like closed photo-bioreactor and open pond technology. Solar light has been used as an energy source in open pond technology and raceway ponds. In municipal WWT autotrophic metabolism removes BOD, nutrient from wastewater in collaboration with heterotrophic bacteria. The pH of this system is high at >9 which leads to the loss of ammonia by volatilization and phosphate by precipitation. Again, the transformation of gaseous CO_2 to liquid is not a feasible process. This type of limitation gives inhibition of biomass growth with a lower pollutant removal rate. In phototrophic culture, the presence of lipid/fatty acid content varied for different algae species.

2. **Heterotrophy:** In heterotrophic culture, light is not necessary, and the energy is obtained from organic chemicals like sugar and organic acids. The energy density of organic chemicals is much higher than CO_2; ΔH for glucose is –2801 kJ/mol, and ΔH for CO_2 is –395 kJ/mol (Nirmalakhandan et al., 2019). As the light is not necessary during culture, this technique gives offers higher biomass density and productivity which can be done in smaller surface-to-volume ratio reactors. Here microalgal cells should have the capability to face organic carbon, inorganic carbon, and nutrients to achieve success (Zhou et al., 2012). The heterotrophic growth of *Chlorella protothecoides* on corn powder leads to 3.4 times higher biomass and 4.2 times higher lipid content than autotrophic growth (Liang et al., 2009). Though the addition of external organic substances makes this culture costlier than autotrophy.

3. **Mixotrophy:** Mixotrophic culture is the combination of photoautotrophic and heterotrophic culture. In this mode, algae acquire their carbon from CO_2 or organic substances and energy from a light source or organic-inorganic chemicals, or both. Studies show that mixotrophic growth rate is much higher than the autotrophic or heterotrophic and sometimes the sum of the two (Nirmalakhandan et al., 2019). In a mixotrophic culture *Scenedesmus* species has been used commonly for nutrient subtraction (Zhou et al., 2012). One of the main advantages of this culture is the production of higher lipid content algal biomass in certain strains. In WWT mixotrophic culture is most appropriate than others.

5.3.2 NUTRIENTS IN ALGAL PONDS

The source of inorganic nitrogen in wastewater includes the use of nitrogenous fertilizers and wastes from humans and animals, which are causative factors of methemoglobinemia and sources of nitrosamines (Tam and Wong, 1989). Orthophosphate is commonly found in urban secondary effluents. Although it does not directly create human health problems, its removal is necessary to reduce eutrophication (Comeau et al., 1987). Nutrients are effectively removed by microalgae in algal ponds due to their high capacity to uptake inorganic nutrients. In this context, species of *Chlorella* and *Scenedesmus* are recognized as important strains for high-performance nutrient removal. *Chlorella vulgaris* removes 86% of inorganic N and 70% of inorganic P with a 2-day retention time when the cells of this unicellular green alga are acclimated to wastewater for 14 days (Lau et al., 1996). However, studies show that these removal percentages may vary depending on the type of wastewater (Wang et al., 2010). The temporal pattern of nitrogen cycling in pond systems was investigated by Rose and coworkers through the analysis of denitrification and anammox genes (Rose et al., 2019). Treatment of wastewater by microalgae leads to the production of several value-added products, such as biofuels, proteins, carbohydrates, and vitamins, from the biomass (Wang et al., 2017). The nutrient removal efficiency (NRE) can be determined using the following formula (Sutherland et al., 2020):

$$\text{NRE}=\frac{\text{(Influent concentration-Effluent concentation)}}{\text{Organic matter}}\times\frac{1}{\text{Retention time}(d)}$$

5.3.3 BIODEGRADATION IN ALGAL PONDS

Microalgae in algal ponds provide an effective pathway to remove organic pollutants such as dyes, crude oil, pesticides, insecticides, xenobiotics, and others through catabolism. In this green approach, unicellular green algae or coenobic algae are particularly important (Ghasemi et al., 2011). Biodegradation occurs mainly through heterotrophic metabolism, extracellular enzymes, or symbiotic relationships between bacteria and microalgae. An increase in pH of the medium, warm temperatures, and high oxygen production during photosynthesis also favor the process (Norvill et al., 2016). Biodegradation of organic compounds under aerobic and anaerobic conditions leads to the formation of simpler chemical structures and may result in complete mineralization (Garcia-Rodríguez et al., 2014; Gruchlik et al., 2018). For biodegradation, species such as *Chlorella*, *Ankistrodesmus*, *Scenedesmus*, and *Pseudomonas* are of significant importance. *Chlorella vulgaris* and *Chlorella pyrenoidosa* successfully degrade *p*-nitrophenol under nonoptimized conditions (Lima et al., 2003). Biodegradation of salicylate occurs rapidly at a rate of 18 mg/L h^{-1} by *Chlorella sorokiniana* due to its rapid growth rate (Muñoz et al., 2003). The single-celled green alga *Chlamydomonas reinhardtii* and the blue-green alga *Anabaena cylindrica* have a strong ability to reduce low concentrations of 2,4-dinitrophenol (Hirooka et al., 2006). Degradation of *p*-nitrophenol and *p*-chlorophenol can be achieved by a microalgal consortium of *Chlorella vulgaris* and *Coenochloris pyrenoidosa* under different photoregimes (Lima et al., 2004). Bisphenol A (BPA) can be metabolized to BPA glycosides by *Pseudokirchneriella subcapitata*, *Scenedesmus acutus*, *Scenedesmus quadricauda*, and *Coelastrum reticulatum*, where the degree of reduction in culture is higher than the degree of accumulation in the cells (Nakajima et al., 2007). The presence of BPA inhibits the growth of *Chlorella sorokiniana* in algal ponds, which can be improved through bacterial growth (Eio et al., 2015). The marine alga *Dunaliella tertiolecta* is capable of biodegrading five model naphthenic acids, such as cyclohexane carboxylic acid, cyclohexane acetic acid, cyclohexane propionic acid, cyclohexane butyric acid, and 1,2,3,4-tetrahydro-2-naphthoic acid, in tailings ponds (Quesnel et al., 2011). In this culture, 300 mg L^{-1} naphthenic acids were used, which is much higher than concentrations typically found in tailings ponds. In antibiotic biodegradation, antibiotics are broken down inside or outside the algal cells by algal enzymes, and the derivatives are subsequently consumed by the cells. Ceftazidime by *Chlorella pyrenoidosa*, levofloxacin by *Chlorella vulgaris*, and *Scenedesmus obliquus* follow intracellular degradation (Leng et al., 2020). Studies show that microalgal-bacterial consortia are able to improve biodegradation compared with single microorganisms (Matamoros et al., 2015).

5.3.4 PHOTODEGRADATION IN ALGAL PONDS

Photodegradation of organic pollutants in algal WWT ponds by microalgae depends on light penetration and its wavelength. Among several ponds, the maturation pond exhibits the highest light penetration capability due to its lower biomass (Curtis et al., 1994). Rühmland et al. (2015) found that the removal efficiency of pharmaceutical compounds decreases with increasing depth, and at 40 cm depth, the compounds are persistent as dark control. The study shows that photo-degradation of micro-pollutants occurs up to 20 cm. Algae release organic molecules (Aota and Nakajima, 2001), radicals, and radical scavengers (Hirayama et al., 1995), which enhance the photo-degradation mechanism. A photosynthetically aerated batch process using microalgae reduces ammonium ion production during the degradation of acetonitrile compared to the mechanically aerated batch process, which becomes much lower in the algal-bacterial process (Munoz and Guieysse, 2006). The antibiotic ciprofloxacin photo-degrades in algal ponds under sunlight irradiation, and the formation of bio-solids can be removed through conventional biological treatment, following pseudo first-order kinetics (Hom-Diaz et al., 2017). For this purpose, HRAPs are the most efficient. HRAPs are particularly designed to optimize light penetration with the biosorption capacity of algal-bacterial biomass that controls the ecological and biological conditions of wastewater to affect the fate of antibiotics (de Godos et al., 2012). The presence of high nitrite concentration in an algal pond inhibits ciprofloxacin photo-degradation due to nitrite acting as a scavenger (Ge et al., 2010). Tetracycline is removed effectively in HRAPs and follows pseudo first-order kinetics, as reported by de Godas et al. (2012). The study shows that in the presence of only solar irradiation (absence of biomass), it leads to 81% removal, but it comes to 50–71% in the presence of biomass only after 43 hours. The removal efficiency reaches 92–98% in the presence of both solar irradiation and biomass, suggesting the combined effect of photodegradation and HRAP biomass. *Coelastrum* sp. removes 64 pharmaceutical and personal care products in HRAPs from municipal wastewater at a rate 5–50% higher than that of the conventional activated sludge process (Villar-Navarro et al., 2018). The degradation efficiency of organic pollutants such as CH_4, CH_3OH, and $C_6H_{12}O_6$ depends on the carbon oxidation-reduction state (CORS) in enclosed algal-bacterial photobioreactors. The lower CORS values of CH_4 (–4) and CH_3OH (–2) lead to a higher oxygen requirement than $C_6H_{12}O_6$ (0), resulting in improved degradation efficiency under photosynthetic oxygenation (Bahr et al., 2011). Stoichiometric analysis shows that, in pollutants, two-thirds of the carbon is associated with bacterial energy formation, whereas one-third is used for biomass buildup (Rocha-Rios et al., 2011). The human metabolites carbamazepine, 2-hydroxycarbamazepine, and 3-hydroxycarbamazepine are degraded under solar irradiation in constructed wetlands (Rühmland et al., 2015). During the photodegradation of antibiotics by microalgae, the generation of reactive species such as hydroxyl radicals, dioxygen radicals, singlet oxygen, and solvated electrons plays an important role; this process is known as indirect photodegradation. In this case, dissolved organic matter, nitrite, and nitrate act as sensitizers to produce reactive species. Antibiotic degradation by direct sunlight absorption leads to bond cleavage, which is known as direct photolysis (Lin and Reinhard, 2005). Tetracycline, ciprofloxacin, cefalexin, and cephradine undergo indirect photodegradation in HRAPs (Leng et al., 2020).

5.4 PARAMETERS AFFECTING ALGAL PONDS

5.4.1 PH

Though the suitable pH range for the growth of microalgae varies from species to species, the majority follow the range between 7 and 9 (Yen et al., 2019). The pH range in algal ponds affects algal growth, photosynthesis, metabolism, the fate of nutrients, and autotrophic and heterotrophic microbial activity (Park et al., 2011). The microalga *Spirulina platensis* is alkalophilic as it prefers a pH range between 9 and 10 for its growth and metabolism, whereas *Chlorococcum littorale* is acidophilic as it prefers a pH of 5 to 6 (Gonçalves et al., 2017). Generally, freshwater microalgae are more tolerant of pH than marine microalgae (Yen et al., 2019). In HRAPs, CO_2 uptake by microalgae leads to an increase in pH (10–11), which is efficient for pathogen removal but decreases pollutant removal capacity. High pH also influences N and P removal (Munoz and Guieysse, 2006).

5.4.2 TEMPERATURE

Algal productivity and algal-based WWT efficiency increase with increasing pond temperature up to an optimum level, which is often 28°C to 35°C (Park et al., 2011). A reported study shows that the removal efficiency of *Chlorella sorokiniana* with *Ralstonia basilensis* strain doubles with increasing temperature from 25°C to 30°C (Munoz and Guieysse, 2006). Removal efficiency at low temperatures below 10°C in HRAP has been studied and described by Grönlund (2004); Munoz and Guieysse (2006). Temperature also affects different parameters in the algal pond, such as ionic equilibrium, pH of the medium, O_2 and CO_2 solubility, and the variation depends on algal species (Bouterfas et al., 2002). High temperatures reduce microalgae production as well as degradation capability due to the denaturation of algal proteins or enzymes (Yen et al., 2019).

5.4.3 LIGHT SUPPLY

Light intensity of 200–400 $\mu Em^{-2}s^{-1}$ is effective for microalgal activity in the algal pond, where higher light intensity causes photoinhibition (Munoz and Guieysse, 2006). Photosynthesis depends on light intensity, which converts inorganic carbon (from CO_2) into organic carbon. Thus, with increasing light intensity, photosynthesis also increases up to the light saturation point in the absence of nutrient limitation. After the saturation point, photoinhibition causes damage to the light receptors of algae, which reduces microalgal activity (Park et al., 2011). Microalgae exhibit high activity at the light saturation point, although this is often impossible due to the light-shading effect caused by high cell concentration or biofilm on the surface of an algal pond in practical cultivation systems. The light-shading effect and photoinhibition can be reduced by mixing the cells (Yen et al., 2019).

5.4.4 DISSOLVED OXYGEN

During the daytime, the dissolved oxygen in HRAPs becomes >200% due to intense photosynthesis by microalgae (Park et al., 2011). High oxygen concentration favors the degradation of organic pollutants and also enhances algal productivity. High dissolved oxygen in water leads to a decrease in photosynthesis O_2 production (Munoz and Guieysse, 2006).

5.5 CONCLUSION

Microalgae-based algal ponds are effective in degrading organic pollutants through a cost-effective, green approach. Algal growth and metabolism depend on the nutrient ratio as well as wastewater characteristics. Therefore, microalgae should be chosen based on the wastewater, the climate of the locality, and the availability of land. Currently, HRAPs are the most effective way for WWT as well as algal biomass production. Separation of waste from the treatment plant is also a typical task that can be improved through membrane PBRs, algae granulation, and biofilm. Improvement of these new technologies requires scaling up and better design guidelines to enhance biomass control and improve filtration.

KEYWORDS

- **biochemical oxygen demand**
- **microalgae**
- **nutrients**
- **photo-autotrophy**
- **photo-bioreactors**
- **photo-degradation**
- **wastewater treatment**

REFERENCES

Abdel-Raouf, N., Al-Homaidan, A. A., & Ibraheem, I. B. M. (2012). Microalgae and wastewater treatment. *Saudi Journal of Biological Sciences, 19*(3), 257–275.

Abeliovich, A., & Weisman, D. (1978). Role of heterotrophic nutrition in growth of the alga *Scenedesmus obliquus* in high-rate oxidation ponds. *Applied and Environmental Microbiology, 35*(1), 32–37.

Acién, F. G., Gómez-Serrano, C., del Morales-Amaral, M. M., Fernández-Sevilla, J. M., & Molina-Grima, E. (2016). Wastewater treatment using microalgae: How realistic a contribution might it be to significant urban wastewater treatment? *Applied Microbiology and Biotechnology, 100*(21), 9013–9022.

Aota, Y., & Nakajima, H. (2001). Mutualistic relationships between phytoplankton and bacteria caused by carbon excretion from phytoplankton. *Ecological Research, 16*(2), 289–299.

Arnold, K. E., Boxall, A. B. A., Brown, A. R., Cuthbert, R. J., Gaw, S., Hutchinson, T. H., et al. (2013). Assessing the exposure risk and impacts of pharmaceuticals in the environment on individuals and ecosystems. *The Royal Society, 9(4), 20130492.*

Aslan, S., & Karapinar Kapdan, I. (2006). Batch kinetics of nitrogen and phosphorus removal from synthetic wastewater by algae. *Ecological Engineering, 28*(1), 64–70.

Baharoglu, Z., & Mazel, D. (2014). SOS, the formidable strategy of bacteria against aggressions. *FEMS Microbiology Reviews, 38*(6), 1126–1145.

Bahr, M., Stams, A. J. M., De la Rosa, F., García-Encina, P. A., & Muñoz, R. (2011). Assessing the influence of the carbon oxidation-reduction state on organic pollutant biodegradation in algal–bacterial photobioreactors. *Applied Microbiology and Biotechnology, 90*(4), 1527–1536.

Berland, J. M., & Cooper, P. F. (2001). *Extensive Wastewater Treatment Processes Adapted to Small and Medium Sized Communities (500 to 5000 population equivalents).* Office of Official Publications of the European Union, Luxembourg. https://op.europa.eu/en/publication-detail/-/publication/a08310d9-83d8-48dd-98fe-25c80b4dc4c2 (accessed on 20th February 2026).

Bouterfas, R., Belkoura, M., & Dauta, A. (2002). Light and temperature effects on the growth rate of three freshwater algae isolated from a eutrophic lake. *Hydrobiological, 489*(1–3), 207–217.

Cai, W., Zhao, Z., Li, D., Lei, Z., Zhang, Z., & Lee, D. J. (2019). Algae granulation for nutrients uptake and algae harvesting during wastewater treatment. *Chemosphere, 214*, 55–59.

Cheng, X., Delanka-Pedige, H. M. K., Munasinghe-Arachchige, S. P., Abeysiriwardana-Arachchige, I. S. A., Smith, G. B., Nirmalakhandan, N., et al. (2020). Removal of antibiotic resistance genes in an algal-based wastewater treatment system employing *Galdieria sulphuraria*: A comparative study. *Science of the Total Environment, 711*, 134435.

Chiaramonti, D., Prussi, M., Casini, D., Tredici, M. R., Rodolfi, L., Bassi, N., et al. (2013). Review of energy balance in raceway ponds for microalgae cultivation: Re-thinking a traditional system is possible. *Applied Energy, 102*, 101–111.

Comeau, Y., Rabionwitz, B., Hall, K. J., & Oldham, W. K. (1987). Phosphate release and uptake in enhanced biological phosphorus removal from wastewater. *Journal of Water Pollution Control Federation, 59(7)*, 707–715.

Craggs, R. J., Lundquist, T. J., & Benemann, J. R. (2013). Wastewater treatment and algal biofuel production. In *Algae for Biofuels and Energy* (pp. 153–163). Springer.

Curtis, T. P., Mara, D. D., Dixo, N. G. H., & Silva, S. A. (1994). Light penetration in waste stabilization ponds. *Water Research, 28*(5), 1031–1038.

Darley, W. M. (1983). Algal biology: A physiological approach. *Phycological, 22*(4), 457–458.

de Godos, I., Muñoz, R., & Guieysse, B. (2012). Tetracycline removal during wastewater treatment in high-rate algal ponds. *Journal of Hazardous Materials, 229*, 446–449.

Delanka-Pedige, H. M. K., Munasinghe-Arachchige, S. P., Cornelius, J., Henkanatte-Gedera, S. M., Tchinda, D., Zhang, Y., et al. (2019). Pathogen reduction in an algal-based wastewater treatment system employing *Galdieria sulphuraria. Algal Research, 39*, 101423.

Ding, C., & He, J. (2010). Effect of antibiotics in the environment on microbial populations. *Applied Microbiology and Biotechnology, 87*(3), 925–941.

Eio, E. J., Kawai, M., Niwa, C., Ito, M., Yamamoto, S., & Toda, T. (2015). Biodegradation of bisphenol A by an algal-bacterial system. *Environmental Science and Pollution Research, 22*(19), 15145–15153.

Gao, P., Munir, M., & Xagoraraki, I. (2012). Correlation of tetracycline and sulfonamide antibiotics with corresponding resistance genes and resistant bacteria in a conventional municipal wastewater treatment plant. *Science of the Total Environment, 421*, 173–183.

García, D., Posadas, E., Grajeda, C., Blanco, S., Martínez-Páramo, S., Acién, G., et al. (2017). Comparative evaluation of piggery wastewater treatment in algal-bacterial photobioreactors under indoor and outdoor conditions. *Bioresource Technology, 245*, 483–490.

Garcia, J., Mujeriego, R., & Hernandez-Marine, M. (2000). High-rate algal pond operating strategies for urban wastewater nitrogen removal. *Journal of Applied Phycology, 12*(3–5), 331–339.

Garcia, J., Mujeriego, R., Bourrouet, A., Penuelas, G., & Freixes, A. (2000). Wastewater treatment by pond systems: Experiences in Catalonia, Spain. *Water Science and Technology, 42*(10, 11), 35–42.

Garcia-Rodríguez, A., Matamoros, V., Fontàs, C., & Salvadó, V. (2014). The ability of biologically based wastewater treatment systems to remove emerging organic contaminants—A review. *Environmental Science and Pollution Research, 21*(20), 11708–11728.

Ge, L., Chen, J., Wei, X., Zhang, S., Qiao, X., Cai, X., et al. (2010). Aquatic photochemistry of fluoroquinolone antibiotics: Kinetics, pathways, and multivariate effects of main water constituents. *Environmental Science & Technology, 44*(7), 2400–2405.

Gentili, F. G., & Fick, J. (2017). Algal cultivation in urban wastewater: An efficient way to reduce pharmaceutical pollutants. *Journal of Applied Phycology, 29*(1), 255–262.

Ghasemi, Y., Rasoul-Amini, S., & Fotooh-Abadi, E. (2011). The biotransformation, biodegradation, and bioremediation of organic compounds by microalgae. *Journal of Phycology, 47*(5), 969–980.

Gonçalves, A. L., Pires, J. C. M., & Simões, M. (2017). A review on the use of microalgal consortia for wastewater treatment. *Algal Research, 24*, 403–415.

González, L. E., Canizares, R. O., & Baena, S. (1997). Efficiency of ammonia and phosphorus removal from a Colombian agro-industrial wastewater by the microalgae *Chlorella vulgaris* and *Scenedesmus dimorphus*. *Bioresource Technology, 60*(3), 259–262.

Grönlund, E. (2004). *Microalgae at Wastewater Pond Treatment in Cold Climate: An Ecological Engineering Approach.* Luleå Tekniska Universitet.

Gruchlik, Y., Linge, K., & Joll, C. (2018). Removal of organic micropollutants in waste stabilization ponds: A review. *Journal of Environmental Management, 206*, 202–214.

Henkanatte-Gedera, S. M., Selvaratnam, T., Caskan, N., Nirmalakhandan, N., Van Voorhies, W., & Lammers, P. J. (2015). Algal-based, single-step treatment of urban wastewaters. *Bioresource Technology, 189*, 273–278.

Hirayama, S., Ueda, R., & Sugata, K. (1995). Effect of hydroxyl radical on intact microalgal photosynthesis. *Energy Conversion and Management, 36*(6–9), 685–688.

Hirooka, T., Nagase, H., Hirata, K., & Miyamoto, K. (2006). Degradation of 2,4-dinitrophenol by a mixed culture of photoautotrophic microorganisms. *Biochemical Engineering Journal, 29*(1, 2), 157–162.

Hoh, D., Watson, S., & Kan, E. (2016). Algal biofilm reactors for integrated wastewater treatment and biofuel production: A review. *Chemical Engineering Journal, 287*, 466–473.

Holmes, B., Paddock, M. B., VanderGheynst, J. S., & Higgins, B. T. (2020). Algal photosynthetic aeration increases the capacity of bacteria to degrade organics in wastewater. *Biotechnology and Bioengineering, 117*(1), 62–72.

Hom-Diaz, A., Norvill, Z. N., Blánquez, P., Vicent, T., & Guieysse, B. (2017). Ciprofloxacin removal during secondary domestic wastewater treatment in high-rate algal ponds. *Chemosphere, 180*, 33–41.

Källqvist, T., & Svenson, A. (2003). Assessment of ammonia toxicity in tests with the microalga *Nephroselmis pyriformis* (Chlorophyta). *Water Research, 37*(3), 477–484.

Kube, M., Jefferson, B., Fan, L., & Roddick, F. (2018). The impact of wastewater characteristics, algal species selection and immobilization on simultaneous nitrogen and phosphorus removal. *Algal Research, 31*, 478–488.

Lau, P. S., Tam, N. F. Y., & Wong, Y. S. (1996). Wastewater nutrients removal by *Chlorella vulgaris*: Optimization through acclimation. *Environmental Technology, 17*(2), 183–189.

Leng, L., Wei, L., Xiong, Q., Xu, S., Li, W., Lv, S., Lu, Q., Wan, L., Wen, Z., & Zhou, W. (2020). Use of microalgae-based technology for the removal of antibiotics from wastewater: A review. *Chemosphere, 238*, 124680.

Li, Y., Chen, Y. F., Chen, P., Min, M., Zhou, W., Martinez, B., et al. (2011). Characterization of a microalga *Chlorella sp.* well adapted to highly concentrated municipal wastewater for nutrient removal and biodiesel production. *Bioresource Technology, 102*(8), 5138–5144.

Liang, Y., Sarkany, N., & Cui, Y. (2009). Biomass and lipid productivities of *Chlorella vulgaris* under autotrophic, heterotrophic and mixotrophic growth conditions. *Biotechnology Letters, 31*(7), 1043–1049.

Lima, S. A. C., Castro, P. M. L., & Morais, R. (2003). Biodegradation of *p*-nitrophenol by microalgae. *Journal of Applied Phycology, 15*(2, 3), 137–142.

Lima, S. A. C., Raposo, M. F. J., Castro, P. M. L., & Morais, R. M. (2004). Biodegradation of *p*-chlorophenol by a microalgae consortium. *Water Research, 38*(1), 97–102.

Lin, A. Y. C., & Reinhard, M. (2005). Photodegradation of common environmental pharmaceuticals and estrogens in river water. *Environmental Toxicology and Chemistry, 24*(6), 1303–1309.

Madakka, M., Jayaraju, N., Rajesh, N., & Subhosh Chandra, M. R. G. (2019). Development in the treatment of municipal and industrial wastewater by microorganism. In *Recent Developments in Applied Microbiology and Biochemistry* (pp. 263–273). Elsevier.

Massoud, M. A., Tarhini, A., & Nasr, J. A. (2009). Decentralized approaches to wastewater treatment and management: Applicability in developing countries. *Journal of Environmental Management, 90*(1), 652–659.

Matamoros, V., Gutiérrez, R., Ferrer, I., García, J., & Bayona, J. M. (2015). Capability of microalgae-based wastewater treatment systems to remove emerging organic contaminants: A pilot-scale study. *Journal of Hazardous Materials, 288*, 34–42.

Muñoz, R., & Guieysse, B. (2006). Algal–bacterial processes for the treatment of hazardous contaminants: A review. *Water Research, 40*(15), 2799–2815.

Muñoz, R., Köllner, C., Guieysse, B., & Mattiasson, B. (2003). Salicylate biodegradation by various algal-bacterial consortia under photosynthetic oxygenation. *Biotechnology Letters, 25*(22), 1905–1911.

Nakajima, N., Teramoto, T., Kasai, F., Sano, T., Tamaoki, M., Aono, M., et al. (2007). Glycosylation of bisphenol A by freshwater microalgae. *Chemosphere, 69*(6), 934–941.

Nirmalakhandan, N., Selvaratnam, T., Henkanatte-Gedera, S. M., Tchinda, D., Abeysiriwardana-Arachchige, I. S. A., Delanka-Pedige, H. M. K., et al. (2019). Algal wastewater treatment: Photoautotrophic vs. mixotrophic processes. *Algal Research, 41*, 101569.

Norvill, Z. N., Shilton, A., & Guieysse, B. (2016). Emerging contaminant degradation and removal in algal wastewater treatment ponds: Identifying the research gaps. *Journal of Hazardous Materials, 313*, 291–309.

Oron, G., Shelef, G., Levi, A., Meydan, A., & Azov, Y. (1979). Algae/bacteria ratio in high-rate ponds used for waste treatment. *Applied and Environmental Microbiology, 38*(4), 570–576.

Oswald, W. J. (1990). Advanced integrated wastewater pond systems. In *Supplying Water and Saving the Environment for Six Billion People* (pp. 73–81). ASCE.

Oswald, W. J., & Golueke, C. G. (1960). Biological transformation of solar energy. In *Advances in Applied Microbiology* (Vol. 2, pp. 223–262). Elsevier.

Park, J. B. K., & Craggs, R. J. (2010). Wastewater treatment and algal production in high-rate algal ponds with carbon dioxide addition. *Water Science and Technology, 61(3)*, 633–639.

Park, J. B. K., Craggs, R. J., & Shilton, A. N. (2011). Wastewater treatment high-rate algal ponds for biofuel production. *Bioresource Technology, 102*(1), 35–42.

Park, J., Jin, H. F., Lim, B. R., Park, K. Y., & Lee, K. (2010). Ammonia removal from anaerobic digestion effluent of livestock waste using green alga *Scenedesmus sp. Bioresource Technology, 101*(22), 8649–8657.

Pinto, G., Pollio, A., Previtera, L., & Temussi, F. (2002). Biodegradation of phenols by microalgae. *Biotechnology Letters, 24*(24), 2047–2051.

Pinto, G., Pollio, A., Previtera, L., Stanzione, M., & Temussi, F. (2003). Removal of low molecular weight phenols from olive oil mill wastewater using microalgae. *Biotechnology Letters, 25*(19), 1657–1659.

Pittman, J. K., Dean, A. P., & Osundeko, O. (2011). The potential of sustainable algal biofuel production using wastewater resources. *Bioresource Technology, 102*(1), 17–25.

Quesnel, D. M., Bhaskar, I. M., Gieg, L. M., & Chua, G. (2011). Naphthenic acid biodegradation by the unicellular alga *Dunaliella tertiolecta. Chemosphere, 84*(4), 504–511.

Rawat, I., Gupta, S. K., Shriwastav, A., Singh, P., Kumari, S., & Bux, F. (2016). Microalgae applications in wastewater treatment. In *Algae Biotechnology* (pp. 249–268). Springer.

Rawat, I., Ranjith Kumar, R., Mutanda, T., & Bux, F. (2011). Dual role of microalgae: Phycoremediation of domestic wastewater and biomass production for sustainable biofuels production. *Applied Energy, 88*(10), 3411–3424.

Rayen, F., Behnam, T., & Dominique, P. (2019). Optimization of a raceway pond system for wastewater treatment: A review. *Critical Reviews in Biotechnology, 39*(3), 422–435.

Renuka, N., Sood, A., Ratha, S. K., Prasanna, R., & Ahluwalia, A. S. (2013). Evaluation of microalgal consortia for treatment of primary treated sewage effluent and biomass production. *Journal of Applied Phycology, 25*(5), 1529–1537.

Rocha-Rios, J., Quijano, G., Thalasso, F., Revah, S., & Muñoz, R. (2011). Methane biodegradation in a two-phase partition internal loop airlift reactor with gas recirculation. *Journal of Chemical Technology and Biotechnology, 86*(3), 353–360.

Rose, A., Munksgaard, N., Kaestli, M., Bodrossy, L., Van de Kamp, J., Tsoukalis, S., et al. (2019). Wastewater nitrogen budgets can be resolved by complementary functional gene and physicochemical methods. *Journal of Water Process Engineering, 29*, 100802.

Rühmland, S., Wick, A., Ternes, T. A., & Barjenbruch, M. (2015). Fate of pharmaceuticals in a subsurface flow constructed wetland and two ponds. *Ecological Engineering, 80*, 125–139.

Salama, E. S., Kurade, M. B., Abou-Shanab, R. A. I., El-Dalatony, M. M., Yang, I. S., Min, B., et al. (2017). Recent progress in microalgal biomass production coupled with wastewater treatment for biofuel generation. *Renewable and Sustainable Energy Reviews, 79*, 1189–1211.

Schwarz, K. R., Sidhu, J. P. S., Toze, S., Li, Y., Lee, E., Gruchlik, Y., et al. (2019). Decay rates of *Escherichia coli, Enterococcus spp.*, F-specific bacteriophage MS2, somatic coliphage and human adenovirus in facultative pond sludge. *Water Research, 154*, 62–71.

Selvaratnam, T., Pegallapati, A. K., Montelya, F., Rodriguez, G., Nirmalakhandan, N., Van Voorhies, W., et al. (2014). Evaluation of a thermo-tolerant acidophilic alga, *Galdieria sulphuraria*, for nutrient removal from urban wastewaters. *Bioresource Technology, 156*, 395–399.

Sonune, A., & Ghate, R. (2004). Developments in wastewater treatment methods. *Desalination, 167*, 55–63.

Suh, I. S., & Lee, C. G. (2003). Photobioreactor engineering: Design and performance. *Biotechnology and Bioprocess Engineering, 8*(6), 313.

Sutherland, D. L., Park, J., Heubeck, S., Ralph, P. J., & Craggs, R. J. (2020). Size matters–Microalgae production and nutrient removal in wastewater treatment high-rate algal ponds of three different sizes. *Algal Research, 45*, 101734.

Tam, N. F. Y., & Wong, Y. S. (1989). Wastewater nutrient removal by *Chlorella pyrenoidosa* and *Scenedesmus sp. Environmental Pollution, 58*(1), 19–34.

Tarlan, E., Dilek, F. B., & Yetis, U. (2002). Effectiveness of algae in the treatment of a wood-based pulp and paper industry wastewater. *Bioresource Technology, 84*(1), 1–5.

Tilley, E., Ulrich, L., Lüthi, C., Reymond, P., & Zurbrügg, C. (2014). *Compendium of Sanitation Systems and Technologies.* Eawag.

Uggetti, E., Sialve, B., Hamelin, J., Bonnafous, A., & Steyer, J. P. (2018). CO_2 addition to increase biomass production and control microalgae species in high-rate algal ponds treating wastewater. *Journal of CO_2 Utilization, 28*, 292–298.

Villar-Navarro, E., Baena-Nogueras, R. M., Paniw, M., Perales, J. A., & Lara-Martín, P. A. (2018). Removal of pharmaceuticals in urban wastewater: High-rate algae pond (HRAP) based technologies as an alternative to activated sludge-based processes. *Water Research, 139*, 19–29.

von Sperling, M. (2007). *Wastewater Characteristics, Treatment and Disposal.* IWA Publishing, London.

Wang, L., Min, M., Li, Y., Chen, P., Chen, Y., Liu, Y., et al. (2010). Cultivation of green algae *Chlorella sp.* in different wastewaters from municipal wastewater treatment plant. *Applied Biochemistry and Biotechnology, 162*(4), 1174–1186.

Wang, M., Keeley, R., Zalivina, N., Halfhide, T., Scott, K., Zhang, Q., et al. (2018). Advances in algal-prokaryotic wastewater treatment: A review of nitrogen transformations, reactor configurations and molecular tools. *Journal of Environmental Management, 217*, 845–857.

Wang, Y., Liu, J., Kang, D., Wu, C., & Wu, Y. (2017). Removal of pharmaceuticals and personal care products from wastewater using algae-based technologies: A review. *Reviews in Environmental Science and Bio/Technology, 16*(4), 717–735.

Xu, M., Li, P., Tang, T., & Hu, Z. (2015). Roles of SRT and HRT of an algal membrane bioreactor system with a tanks-in-series configuration for secondary wastewater effluent polishing. *Ecological Engineering, 85*, 257–264.

Yen, H. W., Hu, I. C., Chen, C. Y., Nagarajan, D., & Chang, J. S. (2019). Design of photobioreactors for algal cultivation. In *Biofuels from Algae* (pp. 225–256). Elsevier.

Young, P., Taylor, M., & Fallowfield, H. J. (2017). Mini review: High-rate algal ponds, flexible systems for sustainable wastewater treatment. *World Journal of Microbiology and Biotechnology, 33*(6), 117.

Zhou, W., Min, M., Li, Y., Hu, B., Ma, X., Cheng, Y., et al. (2012). A hetero-photoautotrophic two-stage cultivation process to improve wastewater nutrient removal and enhance algal lipid accumulation. *Bioresource Technology, 110*, 448–455.

CHAPTER 6

Microalgae in Carbon Dioxide Utilization and Wastewater Treatment for a Sustainable Environment

MUNUSAMY SUBATHRA,[1] NARASIMAN NIRMALA,[2] SELVANANTHAM SHANMUGANATHAM DAWN,[2] and JAYASEELAN ARUN[2,*]

[1]*Department of Biotechnology, Aarupadai Veedu Institute of Technology, Vinayaka Mission's Research Foundation, Paiyanoor, Tamil Nadu, India*

[2]*Center for Waste Management, International Research Center, Sathyabama Institute of Science and Technology, Chennai, Tamil Nadu, India*

*Corresponding author

ABSTRACT

Wastewater generation and greenhouse gas emissions are mainly due to the modernization of livelihoods and related activities. Microalgae play a significant role in the sequestration of carbon dioxide (CO_2) and the uptake of nutrients from wastewater simultaneously. Microalgae are cultivated in various systems, such as open raceway ponds and closed photobioreactors (PBRs). Owing to economic considerations, wastewater type, and operational feasibility, the type of cultivation system is selected. Municipal, grey, whey, and dairy wastewaters are commonly used for microalgae cultivation. However, in the future, the most promising strains need to be identified and enhanced via genetic engineering techniques. Integrated processes help industries achieve sustainability and improved

Environmental Phycology: Climate Change Mitigation and Hazardous Pollutant Cleanup Using Algal Resources. Jeyabalan Sangeetha and Devarajan Thangadurai (Eds.)

DOI: 10.1201/9781779647085-6

economic viability. Harvested biomass can be subjected to various processes, such as extraction, thermochemical conversion, and value-added product recovery. Products such as biofuels, platform chemicals, biopolymers, and animal feed are predominantly produced from algal biomass. Algae are economical and high-yielding biomass resources. Additionally, microalgae-based wastewater treatment (WWT) helps in achieving sustainability goals (SDGs).

6.1 INTRODUCTION

Greenhouse gases comprise 68% carbon dioxide (CO_2) resulting from human activity, industrialization, urbanization, and deforestation (Zhou et al., 2017). Nearly 33.4 Gt of CO_2 is released into the environment by these activities every year (Singh and Dhar, 2019). In recent years, algae-based CO_2 sequestration has attracted more attention than chemical absorption processes. Algal-based systems provide significant support to the carbon capture and storage (CCS) process. Solar energy is used in the photosynthesis process, supplemented by nutrients such as phosphorus and nitrogen from water sources for algal growth. During this process, atmospheric CO_2 is fixed in the algal biomass, resulting in oxygen formation. Microalgae are cultivated in a mixotrophic mode with wastewater and growth medium for enhanced biomass production (Rajan et al., 2019). In this process, COD, ammonia, phosphate, and nitrate are removed to a greater extent. Algae are cultivated in high-rate algal pond (HRAP) systems since they require less input electricity (Garfí et al., 2017). Approximately, algal biomass comprises 50% carbon, and it is estimated that algae can fix 1.83 kg of CO_2 per 1 kg of biomass (Pavlik et al., 2017).

Wastewater management is a global challenge in many countries due to a lack of novel treatment technologies. Wastewater must be treated to avoid environmental pollution, contamination of water bodies, and impacts on living beings. Sustainable and effective technologies for WWT are needed in this urbanized environment (Jayaswal et al., 2018). Algae-based WWT technologies will serve as a better process with economical and ease of operation. In recent years, researchers have shifted their attention towards microalgae-based WWT systems (Almomani et al., 2019; Nagarajan et al., 2019). Microalgae uptake nutrients from wastewater and engage in the photosynthesis process for biomass production. Figure 6.1 explains the overall microalgae technology for wastewater and CO_2 sequestration processes.

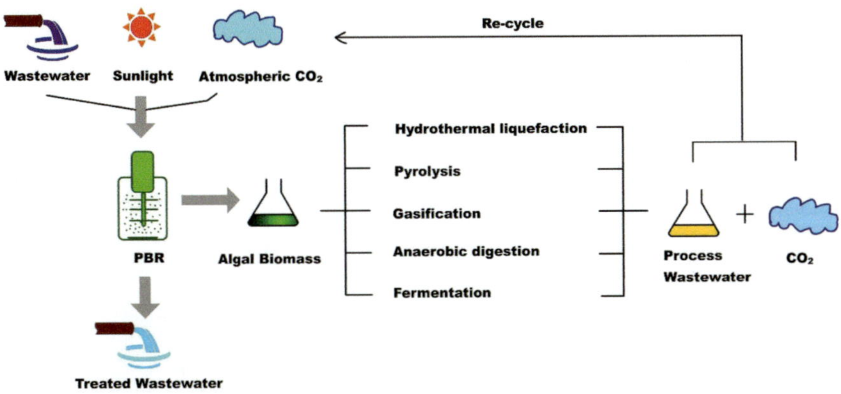

FIGURE 6.1 Microalgae technology for wastewater and CO_2 sequestration.

6.2 WASTEWATER TREATMENT USING MICROALGAE

Anthropogenic activities of industrial development, agricultural practices, and suburbanization pose a threat to the environment universally as they overwhelm it by generating millions of wastewater (Umamaheswari and Shanthakumar, 2016). The continuous disposal of untreated wastewater into the environment leads to pollution and issues such as the degradation of freshwater ecosystems and public health risks. This situation has led to the necessity for effective treatment methods to reduce contaminant concentrations in wastewater before discharging it into natural ecosystems (Abdel-Raouf et al., 2012; Kshirsagar, 2013). Generally, wastewater may contain pathogenic organisms, inorganic compounds, large amounts of oxygen content, and nutrients like nitrogen and phosphorus. The contaminants, mainly nitrogen and phosphorus, are usually removed during the tertiary treatment phase using biological processes, such as anaerobic digestion, chemical methods, nitrification, and denitrification methods. These methods produce enormous amounts of sludge contaminated with other treated chemicals, necessitating further treatment (Gupta et al., 2019; Molazadeh et al., 2019). Therefore, this makes the process more expensive. To overcome the drawbacks associated with tertiary treatment methods, nitrogen and phosphorus-rich industrial wastewater is currently receiving superior consideration for efficient treatment using microalgae (Mohamad et al., 2017). Table 6.1 provides detailed notes on biomass productivity from wastewater.

TABLE 6.1 Literature Survey on Biomass Productivity in Various Carbon Sources

Algae	**Cultivation Medium**	**Carbon Source**	**Specific Growth (Day–1)**	**Biomass Productivity (g/L d)**	**Maximum Biomass Yield (g/L)**	**References**
Scenedesmus obliquus	Municipal wastewater	Air	NA	NA	3.8	Arun et al. (2019)
Scenedesmus abundans	Municipal wastewater	Air	NA	NA	0.003	Rajan et al. (2019)
Spirulina sp.	Distillery wastewater	Air	NA	0.11	NA	Krishnamoorthy et al. (2019)
Chlorella zofingiensis	Kuhl growth medium	CO_2	1.42	2.36	4.43	Zhang et al. (2017)
Chlorella ellipsoidea	Blue-green-11 medium	Air	0.18	0.08	3.2	Purkayastha et al. (2017)
Chlorella sp.	Blue-green-11 medium	CO_2	NA	0.24	1.21	Salati et al. (2017)
Chlorella vulgaris	Aquaculture wastewater	CO_2	0.44	0.04	0.95	Mondal et al. (2017)
Nannochloropsis salina	Walne medium	Glucose	NA	NA	1.85	Marudhupandi et al. (2016)
Keratococcus bicaudatus	Bold's Basal Medium	Air	NA	0.03	NA	Kumar et al. (2016)
Chlorella pyrenoidosa	Blue-green-11 medium	Sucrose	1.02	NA	2.56	Wang et al. (2016)
Scenedesmus obliquus	Urban wastewater	CO_2	0.6	0.20	1.88	Arbib et al. (2014)

Microalgae are considered to be an appropriate resource because of their growth capability in any aqueous medium, their ability to assimilate nutrients from wastewater as a growth supplement, and their eligibility to grow all year round (Nirmala and Dawn, 2016; Satpal and Khambete, 2016; Wollmann et al., 2019). Several studies have attempted to use different wastewaters, such as industrial, refinery, leachate, domestic, and agricultural wastewaters, for microalgal growth. Many trials have been conducted to treat industrial wastewater using potential microalgae to produce biofuel and fix CO_2. Microalgae treatment represents the cultivation of microalgae in nutrient-rich wastewater, thereby utilizing nutrients for their growth and resulting in the harvesting of microalgae for bioenergy production or other value-added products (Gonçalves et al., 2017). Microalgae can efficiently uptake the nitrogen and phosphorus present in the wastewater for their growth. About 80–100% nitrogen and phosphorus utilization efficiency by microalgae has been reported (Arun et al., 2019, 2020). In addition, nutrient removal by microalgae can provide several advantages, such as the production of fertilizers, bioenergy, animal feed, and pharmaceutical purposes.

6.2.1 MICROALGAE CULTIVATION IN PBR

Phototrophic microorganisms are light-harvesting organisms that convert light energy into chemical energy and accumulate different forms of carbon biocompounds (Guihéneuf et al., 2015). Most phototrophic organisms help fix atmospheric CO_2 through photosynthesis. Some of the common light-harvesting microorganisms include cyanobacteria and microalgae, which require light energy and an inorganic carbon source. Therefore, photobioreactors are designed for the effective cultivation of light-harvesting microorganisms, providing an efficient supply of light energy and carbon sources to achieve optimal biomass yield in a confined environment. Third-generation microalgae fuels are primarily utilized due to their advantages, such as cultivation on nonarable land, facilitating the use of wastewater and flue gas for growth, accumulating significant amounts of lipids, and having the potential to produce biodiesel, biohydrogen, bioethanol, biobutanol, syngas, and bioelectricity (Nirmala, 2020; Nirmala et al., 2020). Table 6.2 provides notes on the biomass properties cultivated in wastewater. Large-scale culturing of microalgae species occurs either in open ponds, such as raceway ponds, or in closed systems such as photobioreactors. Open pond systems are relatively economical in terms of construction, maintenance, and operation; however, they face major problems such as contamination, susceptibility to weather conditions, evaporation, and extensive land requirements. In contrast, photobioreactors are cost-effective; however, scaling up microalgae production can be difficult, and there is a risk of cell impairment due to high shear stress conditions in closed systems (Arun et al., 2017). At the same time, the cultivation of phototrophic microorganisms can efficiently eliminate contaminants, reduce medium evaporation, and result in higher biomass concentrations. Currently, there are two major types of cultivation systems: open pond systems and closed cultivation systems. Open pond systems consist of compartments with minimal thickness and are open to the atmosphere. Other types of open pond cultivation systems include circular ponds, unmixed open ponds, and thin-layer inclined ponds. Closed systems, or photobioreactors, are designed as transparent containers for microalgae growth. The different types of photobioreactors include vertical bubble column photobioreactors, horizontal photobioreactors, bubble column photobioreactors, tubular photobioreactors, helical photobioreactors, flat-plate photobioreactors, and combinations of bubble column and inclined tubular photobioreactors. PBRs are mostly designed in the form of pipes, tubes, bags, or plates, which are transparent in nature. Microalgal cultivation is carried out in the presence of light, nutrients, and CO_2. Among the various PBR designs, three widely used methods are discussed below (Carvalho et al., 2006).

6.2.1.1 ANNULAR PBR

The annular PBR is widely used as an air lift reactor or bubble column reactor. They are also rarely used as a stir tank reactor. They are arranged with vertical transparent columns. They are provided with aeration from the bottom of the reactor, and light is supplied with the help of transparent columns. They provide equal mixing, controlled parameters, and high volumetric transfer rates (Eriksen et al., 2007; Lee et al., 2006).

6.2.1.2 TUBULAR PBR

The columns are long and transparent in nature in tubular PBR. The algal biomass is supplied into the PBR by using mechanical pumps or by creating the pumping force through airlift. Airlift is also used to feed the CO_2 for the growing algal biomass (Molina et al., 2001; Travieso et al., 2001).

TABLE 6.2 Literature Survey on Various Algal Biomass Compositions

Algal Species	Algae Growth Medium Source	Elemental Analysis (%)					Proximate Analysis (wt %)			References
		C	H	O	N	S	Moisture	Ash	Volatile Solids	
Chlorella vulgaris	Anaerobic digestion effluent	29.26	3.78	28.26	–	–	–	38.16	55.65	Sztancs et al. (2020)
Scenedesmus abundans	Domestic wastewater	52.26	7.38	31.73	7.53	0.5	2.36	9.82	–	Sundar Rajan et al. (2020)
Scenedesmus obliquus	Municipal wastewater: BG-11 medium (80:20)	–	–	–	–	–	10.7	11.5	–	Arun et al. (2019)
Nannochloropsis sp.	Aquarium water	48.41	9.01	33.91	7.38	1.29	4.10	8.28	82.28	Ebadi and Hisoriev (2019)
Stigonematales	Rainwater	41.28	3.15	41.8	7.43	0.16	4.5	39.5	35	Hossain et al. (2019)
Nannochloropsis salina	F/2 culture medium	44.5	7.3	25.4	9.5	3.0	–	–	–	Cheng et al. (2018)
Gracilaria gracilis	–	36.75	5.86	17.51	2.88	1.99	5.88	36	53.1	Parsa et al. (2018)
Chlamydomonas reinhardtii	Tris-acetate-phosphate (TAP) medium	41.23	7.21	41.3	8.81	1.45	–	17.96	70.83	Andrade et al. (2018)
Oscillatoria sp.	–	37.7	5.5	38.34	4.7	0.46	4.4	8.9	72.3	Kosowska-Golachowska et al. (2017)
Cladophora glomerata	–	43.45	9.17	38.45	0.83	1.56	90	6.54	75.55	Zaini et al. (2017)

6.2.1.3 FLAT PANEL PBR

This is the most efficient method, where flat transparent panels are constructed for the cultivation of algal biomass. The algae cultivated here are denser in nature, and this PBR promotes higher photosynthetic activity by allowing the incoming light to be absorbed by the top layer of the culture (Richmond et al., 2003).

6.2.2 MAJOR FACTORS DISTURBING PHOTO-BIOREACTOR EFFICIENCY

The overall production cost can be reduced by optimizing the culture conditions using the suitable photo-bioreactor. In order to achieve the economic feasibility of microalgae, a superior appreciation of both appropriate photo-bioreactor design and physiological necessities for microalgae growth is essential. Major factors that are liable to impact the efficiency of photo-bioreactor operation and biomass production of microalgae are pH, CO_2, light, temperature, oxygen removal, and agitation (Doucha et al., 2005). Among these factors, light and CO_2 play a major role in photosynthesis; any variation in these two factors affects the photosynthetic efficiency of microalgae, which in turn upsets biomass production. Meanwhile, pH and temperature indirectly affect the growth of microalgae if the optimum conditions are changed. As the proportions of the photo-bioreactor structure for biomass production increase, challenges such as self-shading, high energy requirements, photo-inhibition, sustainable sources of nutrients for microalgae growth, and assurance of concentrated and uncontaminated CO_2 availability become more challenging for the commercialization of biomass production.

6.2.3 MICROALGAE HARVESTING

Harvesting of microalgae is an important downstream processing technique that involves two steps: (i) harvesting and (ii) thickening. These two steps are carried out using flocculation, centrifugation, membrane filtration, flotation, and magnetic separation or a combination of these techniques.

6.2.3.1 FLOCCULATION

The flocculation process is carried out to separate the microalgal cells from the culture media. Since microalgae are less dense in nature, they float on the media. A positive electrode is fixed in the tank along with a cationic substance that acts as a flocculant and forms algal clogs with one another, thereby promoting the separation process. The widely used flocculants are chemical in nature, which include organic salts. This has the disadvantage of modifying the microalgae's nature by altering the amount of starch, lipid, and CO_2 in it. This alteration leads to a lesser yield of biofuel. To overcome this, natural flocculants came into existence, which are costlier than chemical flocculants but do not affect the microalgae in any way. To cut down the cost spent on natural flocculants, bio-flocculants emerged, providing both the advantages of chemical and natural flocculants.

6.2.3.2 CENTRIFUGATION

This separation process uses centrifugal force to separate the algal biomass. The separation efficiency is purely based on the size of the microalgae participating in the process. There are different types of centrifuges, which include hydrocyclones, solid ejecting types, disc centrifuges, nozzle-type centrifuges, solid retaining disc centrifuges, decanters, imperforate basket centrifuges, multi-chambered centrifuges, and tubular centrifuges. This method has the advantage of not using any chemicals for the separation process, thereby not causing any chemical damage to the microalgal cells. The disadvantage is that when the scaling-up process is carried out for the culturing mechanism, the estimated cost for centrifugation is very high due to high energy consumption. Additionally, shear stress and an increase in gravitational force may cause damage to the microalgal cells (Lu et al., 2020).

6.2.3.3 MEMBRANE FILTRATION

Membrane filtration is a physical and mechanical way of separating the solid particles from the liquid by using the membrane of different pore size. Different types of membrane filtration used are forward osmosis, tangential flow filtration, vacuum filtration, pressure filtration, ultrafiltration, microfiltration, cross flow filtration and dead-end filtration. This holds an advantage that, this can be scaled up with very low cost and without the usage chemicals, which may contaminate the entire growing microalgal cells. This possesses a disadvantage of pore blockage, sedimentation of solid over the membrane, thereby decreasing the filtration efficiency, formation of gel like layer, and so forth. To overcome this limitation vibrating membrane were developed, addition of surfactant came into practice, which pawed the way to overcome fouling process over the membrane.

6.2.3.4 FLOTATION

Flotation is the reverse technique of the sedimentation process. Here, air bubbles are used to raise the destabilized particles and make them float on the surface of the culture medium, from which they can be removed by a skimming process. In general, flotation is of four types: (i) dissolved air flotation; (ii) ozonation–dispersed flotation; (iii) electroflotation; and (iv) dispersed air flotation. This is the most efficient way of harvesting microalgae on a large scale if the coagulant used in flotation is replaced by a nontoxic, low-cost, and natural coagulant, since the use of chemical coagulants may cause damage to the microalgal cell surface or cause contamination. In addition, parameters such as pH, coagulant concentration, and flotation rate must be optimized to achieve high efficiency (Lu et al., 2020).

6.2.3.5 MAGNETIC SEPARATION

This process is carried out with a magnetic substance or by sing magnetic nanoparticles. The microalgal cells can be harvested by magnetic separation method, since the microalgal biomass possess negative charge and magnet possess positive charge. This method seemed the attention of the consumers, since it does not have any energy consumption, cost effective and it can be reused. But this method is not adopted due to the low efficiency.

6.3 CARBON DIOXIDE FOR MICROALGAE CULTIVATION

Microalgal growth is highly influenced by the percentage of CO_2 supplied to the culturing medium. The carbon sources can be efficiently utilized by microalgae by converting it into the useful products by carrying out photosynthesis process. Different microalgae possess different tolerance capacity of CO_2, which depends on their absorption capacity. The microalgae cultivation has the ability to remove the green-house gases from various sources, which include – industrial effluent gases, fossil fuel emission and other carbon emissions. In closed PBR the CO_2 gases can be supplied in the form of airlift pumps to the algal culture. By increasing the retention time of the CO_2 inside the reactor, the absorption capacity can also be increased. As the absorption percentage increases, the conversion of CO_2 into starch, lipids and other economically useful products also get increased. Such microalgae can result in high production of CH_4 gas and biofuel production. The highest-yield-producing microalgae are *Haematococcus pluvialis*, *Spirulina* sp., *Chlorella* sp., *Chlorella vulgaris*, *Scenedesmus incrassatulus*, *Scenedesmus obliquus*, *Scenedesmus dimorphus*, and *Scenedesmus* sp., which are identified as promising microalgal strains capable of assimilating higher levels of CO_2 and reducing greenhouse gas emissions from coal power plants, fossil fuel emissions, and other carbon-intensive industrial processes (Chakrabarti et al., 2018).

6.3.1 CO_2 FROM FLUE GAS

Even though microalgae have the same drawbacks as conventional CO_2 capture and storage systems, they are capable of biological CO_2 fixation from flue gas, resulting in the production of value-added biomass. In recent years, major attention has been given to CO_2 reduction due to the increase of greenhouse gases in the atmosphere (Dantas et al., 2012; Wang et al., 2008). A large amount of CO_2 is released into the atmosphere during the burning of oil, natural gas, coal, and chemical processing (Guihéneuf et al., 2015). There are three different sources of CO_2 used for cultivating microalgae: commercial CO_2, flue gas, and atmospheric CO_2. Utilization of commercial CO_2 increases the production cost of microalgae cultivation; at the same time, it decreases the potential of microalgae to mitigate CO_2 levels. The sufficient amount of CO_2 present at any source is also important because microalgae secrete the RuBisCO enzyme in the presence of light, which is responsible for the consumption of CO_2 and acts as a biological carbon sink, converting CO_2 into 3C-phosphoglycerate (3-PGA) molecules in the Calvin cycle. In the case of inadequate CO_2 availability, the RuBisCO enzyme binds to oxygen instead of CO_2, causing photorespiration (Songolzadeh et al., 2014).

Flue gas emission from urban and industrial areas will also contain 5–20% of CO_2. Therefore, these emitted flue gas containing CO_2 can be utilized for biomass production. Flue gas composition is depending on the fuel type and the combustion system (Zheng et al., 2016). As per the US DOE in 2010, the concentration of CO_2 content present power plants was around 5–6% whereas in natural gas ranges from 10–15%. Randomly flue gas emission may consist of 3–25% of CO_2 in it. Flue gas consist of high percentage of CO_2, NOx, sulfur oxide concentration and unburned particles (He and Hägg, 2014). The direct usage of flue gas for microalgae growth will cause extreme difficulties to enhance their growth (Zhang, 2015). Therefore, the flue gases have to be desulfurized to below 60 ppm and cooled before it is introduced into the cultivation system (Mustafa et al., 2016). CO_2 separation from flue gas can be assessed by using membranes, cryogenic fractionation, molecular sieves, calcium oxides, hydrotalcites, carbon nanotubes, lithium zirconate-activated carbon. The recovery of CO_2 process using these adsorbents by performing swinging pressure for desorption technique and swinging temperature for adsorption technique.

6.4 CONCLUSION

In recent studies, microalgae-based WWT systems were used for simultaneous WWT and biomass production for biofuels. Recent studies suggest that *Scenedesmus obliquus* removes 80.19% of COD, 90.73% of nitrate, and 86.31% of phosphate from domestic wastewater. Microalgae biomass yield was around 3.55 g/L when cultivated in PBR. From a broad review of the literature, it is clear that CO_2 from flue gas is purged into wastewater for microalgae cultivation. Biomass production depends on the amount of nutrients in the wastewater and the amount of CO_2 dispersed in the system. The biomass results in a minimum of 50% carbon content and enhanced CO_2 fixation. Microalgae-based WWT systems are effective, economical, and provide a sustainable pathway for waste management.

KEYWORDS

- **carbon dioxide sequestration**
- **greenhouse gas emissions**
- **magnetic nanoparticles**
- **microalgae**
- **photo-bioreactor**
- **phototrophic microorganisms**
- **wastewater treatment**

REFERENCES

Abdel-Raouf, N., Al-Homaidan, A. A., & Ibraheem, I. B. M. (2012). Microalgae and wastewater treatment. *Saudi Journal of Biological Sciences, 19*(3), 257–275.

Almomani, F., Al Ketife, A., Judd, S., Shurair, M., Bhosale, R. R., Znad, H., et al. (2019). Impact of CO_2 concentration and ambient conditions on microalgal growth and nutrient removal from wastewater by a photobioreactor. *Science of the Total Environment, 662*, 662–671.

Andrade, L. A., Batista, F. R. X., Lira, T. S., Barrozo, M. A. S., & Vieira, L. G. M. (2018). Characterization and product formation during the catalytic and non-catalytic pyrolysis of the green microalgae *Chlamydomonas reinhardtii*. *Renewable Energy, 119*, 731–740.

Arbib, Z., Ruiz, J., Álvarez-Díaz, P., Garrido-Pérez, C., & Perales, J. A. (2014). Capability of different microalgae species for phytoremediation processes: Wastewater tertiary treatment, CO_2 bio-fixation and low-cost biofuels production. *Water Research, 49*, 465–474. https://doi.org/10.1016/j.watres.2013.10.036 (accessed on 20th February 2026).

Arun, J., Gopinath, K. P., Shreekanth, S. J., Sahana, R., Raghavi, M. S., & Gnanaprakash, D. (2019). Effects of process parameters on hydrothermal liquefaction of microalgae biomass grown in municipal wastewater. *Petroleum Chemistry, 59*, 194–200.

Arun, J., Gopinath, K. P., Vo, D. V. N., SundarRajan, P., & Swathi, M. (2020). Co-hydrothermal gasification of *Scenedesmus sp.* with sewage sludge for bio-hydrogen production using a novel solid catalyst derived from carbon-zinc battery waste. *Bioresource Technology Reports, 11*, 100459.

Arun, J., Shreekanth, S. J., Sahana, R., Raghavi, M. S., Gopinath, K. P., & Gnanaprakash, D. (2017). Studies on influence of process parameters on hydrothermal catalytic liquefaction of microalgae (*Chlorella vulgaris*) biomass grown in wastewater. *Bioresource Technology, 244*, 963–968.

Carvalho, A. P., Meireles, L. A., & Malcata, F. X. (2006). Microalgal reactors: A review of enclosed system designs and performances. *Biotechnology Progress, 22*, 1490–1506.

Chakrabarti, T., Krishnamurthi, K., Devi, S. S., & Fulke, B. A. (2018). CO_2 sequestration by microalgae: Advances and perspectives. In *OMICS Group eBooks*. OMICS Group.

Cheng, F., Cui, Z., Mallick, K., Nirmalakhandan, N., & Brewer, C. E. (2018). Hydrothermal liquefaction of high- and low-lipid algae: Mass and energy balances. *Bioresource Technology, 258*, 158–167. https://doi.org/10.1016/j.biortech.2018.02.100 (accessed on 20th February 2026).

Dantas, T. L., Rodrigues, A. E., & Moreira, R. (2012). Separation of carbon dioxide from flue gas using adsorption on porous solids. In *Greenhouse Gases: Capturing, Utilization and Reduction* (pp. 57–80).

Doucha, J., Straka, F., & Lívanský, K. (2005). Utilization of flue gas for cultivation of microalgae *Chlorella sp.* in an outdoor open thin layer photobioreactor. *Journal of Applied Phycology, 17*, 403–412.

Ebadi, A. G., & Hisoriev, H. (2019). Gasification of algal biomass (*Cladophora glomerata* L.) with $CO_2/H_2O/O_2$ in a circulating fluidized bed. *Environmental Technology, 40*, 749–755.

Eriksen, N. T., Riisgård, F. K., Gunther, W. S., & Lønsmann Iversen, J. J. (2007). On-line estimation of O_2 production, CO_2 uptake, and growth kinetics of microalgal cultures in a gas-tight photobioreactor. *Journal of Applied Phycology, 19*, 161–174.

Garfí, M., Flores, L., & Ferrer, I. (2017). Life cycle assessment of wastewater treatment systems for small communities: Activated sludge, constructed wetlands and high-rate algal ponds. *Journal of Cleaner Production, 161*, 211–219.

Gonçalves, A. L., Pires, J. C. M., & Simões, M. (2017). A review on the use of microalgal consortia for wastewater treatment. *Algal Research, 24*, 403–415.

Guihéneuf, F., Schmid, M., & Stengel, D. B. (2015). Lipids and fatty acids in algae: Extraction, fractionation into lipid classes, and analysis by gas chromatography coupled with flame ionization detector (GC-FID). In D. B. Stengel & S. Connan (Eds.), *Natural Products from Marine Algae* (pp. 173–190). Springer.

Gupta, S., Pawar, S. B., & Pandey, R. A. (2019). Current practices and challenges in using microalgae for treatment of nutrient-rich wastewater from agro-based industries. *Science of the Total Environment, 687*, 1107–1126.

He, X., & Hägg, M. B. (2014). Energy-efficient process for CO_2 capture from flue gas with novel fixed-site-carrier membranes. *Energy Procedia, 63*, 174–185.

Hossain, N., Zaini, J., Mahlia, T. M. I., & Azad, A. K. (2019). Elemental, morphological and thermal analysis of mixed microalgae species from drain water. *Renewable Energy, 131*, 617–624.

Jayaswal, K., Sahu, V., & Gurjar, B. R. (2018). Water pollution, human health and remediation. In S. Bhattacharya, A. B. Gupta, A. Gupta, & A. Pandey (Eds.), *Water Remediation* (pp. 11–27). Springer.

Kosowska-Golachowska, M., Musiał, T., Urbaniak, D., & Otwinowski, H. (2017). Analysis of microalgae pellets combustion in a circulating fluidized bed. In *E3S Web of Conferences* (p. 2035). EDP Sciences.

Krishnamoorthy, S., Manickam, P., & Muthukaruppan, V. (2019). Evaluation of distillery wastewater treatability in a customized photobioreactor using blue-green microalgae—Laboratory and outdoor study. *Journal of Environmental Management, 234*, 412–423.

Kshirsagar, A. D. (2013). Bioremediation of wastewater by using microalgae: An experimental study. *International Journal of Life Science Biotechnology and Pharma Research, 2*, 339–346.

Kumar, K. S., Prasanthkumar, S., & Ray, J. G. (2016). Experimental assessment of productivity, oil-yield and oil-profile of eight different common freshwater-blooming green algae of Kerala. *Biocatalysis and Agricultural Biotechnology, 8*, 270–277.

Lee, H. S., Seo, M. W., Kim, Z. H., & Lee, C. G. (2006). Determining the best specific light uptake rates for the lumostatic cultures in bubble column photobioreactors. *Enzyme and Microbial Technology, 39*, 447–452.

Lu, W., Alam, M. A., Liu, S., Xu, J., & Parra Saldivar, R. (2020). Critical processes and variables in microalgae biomass production coupled with bioremediation of nutrients and CO_2 from livestock farms: A review. *Science of the Total Environment, 716*, 13524.

Marudhupandi, T., Sathishkumar, R., & Kumar, T. T. A. (2016). Heterotrophic cultivation of *Nannochloropsis salina* for enhancing biomass and lipid production. *Biotechnology Reports, 10*, 8–16. https://doi.org/10.1016/j.btre.2016.02.001 (accessed on 20th February 2026).

Mohamad, S., Fares, A., Judd, S., Bhosale, R., Kumar, A., Gosh, U., et al. (2017). Advanced wastewater treatment using microalgae: Effect of temperature on removal of nutrients and organic carbon. In *IOP Conference Series: Earth and Environmental Science* (p. 012032). IOP Publishing.

Molazadeh, M., Ahmadzadeh, H., Pourianfar, H. R., Lyon, S., & Rampelotto, P. H. (2019). The use of microalgae for coupling wastewater treatment with CO_2 bio fixation. *Frontiers in Bioengineering and Biotechnology, 7*, 42.

Molina, E., Fernández, J., Acién, F. G., & Chisti, Y. (2001). Tubular photobioreactor design for algal cultures. *Journal of Biotechnology, 92*, 113–131.

Mondal, M., Ghosh, A., Tiwari, O. N., Gayen, K., Das, P., Mandal, M. K., et al. (2017). Influence of carbon sources and light intensity on biomass and lipid production of *Chlorella sorokiniana* BTA 9031 isolated from coalfield under various nutritional modes. *Energy Conversion and Management, 145*, 247–254.

Mustafa, J., Farhan, M., & Hussain, M. (2016). CO_2 separation from flue gases using different types of membranes. *Journal of Membrane Science and Technology, 6*, 153–159.

Nagarajan, D., Kusmayadi, A., Yen, H. W., Dong, C. D., Lee, D. J., & Chang, J. S. (2019). Current advances in biological swine wastewater treatment using microalgae-based processes. *Bioresource Technology, 289*, 121718.

Nirmala, N. (2020). Phylogenetic analysis for identification of lipid-enriched microalgae and optimization of extraction conditions for biodiesel production using response surface methodology tool. *Biocatalysis and Agricultural Biotechnology, 25*, 101603.

Nirmala, N., & Dawn, S. S. (2016). Environmental stress control to enhance lipid content in oleaginous microalgae for biodiesel production. *International Journal of Chemtech Research, 9*, 237–241.

Nirmala, N., Dawn, S. S., & Harindra, C. (2020). Analysis of performance and emission characteristics of waste cooking oil and *Chlorella variabilis* MK039712 biodiesel blends in a single-cylinder, four-stroke diesel engine. *Renewable Energy, 147*, 284–292.

Parsa, M., Jalilzadeh, H., Pazoki, M., Ghasemzadeh, R., & Abduli, M. (2018). Hydrothermal liquefaction of *Gracilaria gracilis* and *Cladophora glomerata* macroalgae for biocrude production. *Bioresource Technology, 250*, 26–34. https://doi.org/10.1016/j.biortech.2017.10.059 (accessed on 20th February 2026).

Pavlik, D., Zhong, Y., Daiek, C., Liao, W., Morgan, R., Clary, W., et al. (2017). Microalgae cultivation for carbon dioxide sequestration and protein production using a high-efficiency photobioreactor system. *Algal Research, 25*, 413–420.

Purkayastha, J., Bora, A., Gogoi, H. K., & Singh, L. (2017). Growth of high oil-yielding green alga *Chlorella ellipsoidea* in diverse autotrophic media and effect on its constituents. *Algal Research, 21*, 81–88.

Rajan, P. S., Gopinath, K. P., Arun, J., & Pavithra, K. G. (2019). Hydrothermal liquefaction of *Scenedesmus abundans* biomass spent for sorption of petroleum residues from wastewater and studies on recycling of post-hydrothermal liquefaction wastewater. *Bioresource Technology, 283*, 36–44.

Richmond, A., Cheng-Wu, Z., & Zarmi, Y. (2003). Efficient use of strong light for high photosynthetic productivity: Interrelationships between the optical path, the optimal population density and cell-growth inhibition. *Biomolecular Engineering, 20*, 229–236.

Salati, S., D'Imporzano, G., Menin, B., Veronesi, D., Scaglia, B., Abbruscato, P., et al. (2017). Mixotrophic cultivation of *Chlorella* for local protein production using agro-food by-products. *Bioresource Technology, 230*, 82–89.

Satpal, K. A. K., & Khambete, A. K. (2016). Wastewater treatment using micro-algae: A review paper. *International Journal of Engineering, Technology, Management and Applied Sciences, 4*, 188–192.

Singh, J., & Dhar, D. W. (2019). Overview of carbon capture technology: Microalgal biorefinery concept and state-of-the-art. *Frontiers in Marine Science, 6*, 29.

Songolzadeh, M., Soleimani, M., Takht Ravanchi, M., & Songolzadeh, R. (2014). Carbon dioxide separation from flue gases: A technological review emphasizing reduction in greenhouse gas emissions. *The Scientific World Journal, 2014*, 1–34.

SundarRajan, P., Gopinath, K. P., Arun, J., GracePavithra, K., Pavendan, K., & AdithyaJoseph, A. (2020). An insight into carbon balance of product streams from hydrothermal liquefaction of *Scenedesmus abundans* biomass. *Renewable Energy, 151*, 79–87.

Sztancs, G., Juhasz, L., Nagy, B. J., Nemeth, A., Selim, A., Andre, A., et al (2020). Co-hydrothermal gasification of *Chlorella vulgaris* and hydrochar: The effects of waste-to-solid biofuel production and blending concentration on biogas generation. *Bioresource Technology, 302*, 122793.

Travieso, L., Hall, D. O., Rao, K. K., Benítez, F., Sánchez, E., & Borja, R. (2001). A helical tubular photobioreactor producing *Spirulina* in a semicontinuous mode. *International Biodeterioration & Biodegradation, 47*, 151–155. https://doi.org/10.1016/S0964-8305(01)00043-9 (accessed on 20th February 2026).

Umamaheswari, J., & Shanthakumar, S. (2016). Efficacy of microalgae for industrial wastewater treatment: A review on operating conditions, treatment efficiency and biomass productivity. *Reviews in Environmental Science and Bio/Technology, 15*, 265–284.

Wang, B., Li, Y., Wu, N., & Lan, C. Q. (2008). CO_2 bio-mitigation using microalgae. *Applied Microbiology and Biotechnology, 79*, 707–718.

Wang, S., Wu, Y., & Wang, X. (2016). Heterotrophic cultivation of *Chlorella pyrenoidosa* using sucrose as the sole carbon source by co-culture with *Rhodotorula glutinis*. *Bioresource Technology, 220*, 615–620.

Wollmann, F., Dietze, S., Ackermann, J. U., Bley, T., Walther, T., Steingroewer, J., et al. (2019). Microalgae wastewater treatment: Biological and technological approaches. *Engineering in Life Sciences, 19*, 860–871.

Zaini, I. N., Nurdiawati, A., & Aziz, M. (2017). Cogeneration of power and H_2 by steam gasification and syngas chemical looping of macroalgae. *Applied Energy, 207*, 134–145.

Zhang, X. (2015). *Microalgae Removal of CO_2 from Flue Gas.* IEA Clean Coal Centre, UK.

Zhang, Z., Sun, D., Wu, T., Li, Y., Lee, Y., Liu, J., et al. (2017). The synergistic energy and carbon metabolism under mixotrophic cultivation reveals the coordination between photosynthesis and aerobic respiration in *Chlorella zofingiensis*. *Algal Research, 25*, 109–116.

Zheng, Q., Martin, G. J. O., & Kentish, S. E. (2016). Energy-efficient transfer of carbon dioxide from flue gases to micro-algal systems. *Energy & Environmental Science, 9*, 1074–1082.

Zhou, W., Wang, J., Chen, P., Ji, C., Kang, Q., Lu, B., et al. (2017). Bio-mitigation of carbon dioxide using microalgal systems: Advances and perspectives. *Renewable and Sustainable Energy Reviews, 76*, 1163–1175.

CHAPTER 7

Applications of Omic Tools (Genomics, Transcriptomics, Proteomics, and Metabolomics) Based on Phytoplankton Communities to Assess the Environmental Quality of Marine Systems

SILVIA LOMARTIRE,[1] SARA GARCÍA-POZA,[2] DIANA PACHECO,[2] JOÃO CARLOS MARQUES,[2] LEONEL PEREIRA,[1,3] and ANA MARTA MENDES GONÇALVES[1,3,*]

[1]*CFE—Centre for Functional Ecology: Science for People & the Planet, Associate Laboratory TERRA, Department of Life Sciences, University of Coimbra, Coimbra, Portugal*

[2]*Department of Life Sciences, MARE–Marine and Environmental Sciences Centre/ARNET–Aquatic Research Network, University of Coimbra, Coimbra, Portugal*

[3]*Centre for Functional Ecology – Science for People & the Planet (CFE), Associate Laboratory TERRA, University of Coimbra Campus at Figueira da Foz, Quinta das Olaias, Figueira da Foz, Portugal*

*Corresponding author

ABSTRACT

The present work focuses on the effects of climate change on the environment and marine organisms. In the last decade, the effects of global warming have intensified, causing several problems along trophic webs,

Environmental Phycology: Climate Change Mitigation and Hazardous Pollutant Cleanup Using Algal Resources. Jeyabalan Sangeetha and Devarajan Thangadurai (Eds.)

DOI: 10.1201/9781779647085-7

starting from microalgae. The factors that influence microalgae include temperature rise, ocean acidification (OA), changes in salinity, and increases in CO_2. All these threats cause problems for microalgae and other organisms that consume or utilize phytoplankton; as a result, there is a decline in marine organisms within the food chain. It is important to consider the diseases caused by microalgae (toxic to humans) that spread through harmful algal blooms (HABs). Some species are acclimated to environmental changes; however, some species of microalgae are endangered. Therefore, it is important to protect them for their multiple uses, such as cleanup and bioremediation, as well as for pharmaceutical and nutritional purposes. Over the years, omics tools related to microalgae have been developed, including genomics, transcriptomics, proteomics, and metabolomics, which are useful for predicting changes in the community due to climate change and for forecasting the formation of HABs.

7.1 INTRODUCTION

The ocean is involved in the frontline of global warming changes; it can absorb approximately 93% of the heat generated by greenhouse gas emissions, but it can also absorb CO_2 released into the atmosphere from anthropogenic sources. As a result, the prediction is an increase in sea surface temperature of 1–10°C over the next 100 years (IPCC, 2013) and a significant change in carbonate chemistry, as well as an increased acidity of seawater (IPCC, 2013) to 100–150% compared to collected data from the pre-industrial era (Baltar et al., 2019; IPCC, 2013). Consequently, marine organisms are influenced by these environmental changes, which could lead to the collapse of ecosystems.

The ocean and its ecosystems are a fundamental source of services, provided by the abundance of marine organisms present on our planet. Unfortunately, climate change poses a threat to the abundance and distribution of species (Doney et al., 2012; García Molinos et al., 2016; Pinsky et al., 2013; Pörtner et al., 2014), which may translate into changes in ecosystem structure, functioning, seafood supply, and conservation of marine species (Blanchard et al., 2017; Cheung et al., 2010; Stock et al., 2017).

The Intergovernmental Panel on Climate Change (IPCC) and the Intergovernmental Science-Policy Platform on Biodiversity and Ecosystem Services (IPBES) have developed tools to quantify future global projections (Pörtner et al., 2014) regarding the influence of global warming on the ocean and marine organisms. In recent years, a large number of marine ecosystem models (MEMs) have been developed (Tittensor et al., 2018) to understand trends in species distribution, biomass, and fishery catch (Barange et al., 2014; Blanchard et al., 2012; Cheung et al., 2010; Galbraith et al., 2017; Lefort et al., 2015). However, it remains unclear how reliable these models are based on projections of climate change scenarios (Lotze et al., 2019). For example, changes in species distribution are already assumed to be consequences of climate change (Free et al., 2019), but unusual migration alone cannot be the only response to environmental changes (Estes et al., 2016). Therefore, predictions of climate change remain a contentious issue; nonetheless, many studies show that climate change affects all organisms through the trophic web (Sommer and Lengfellner, 2008).

The role of microalgae is crucial as a source of primary organic matter for zooplankton (Kang et al., 2003). Therefore, it is possible to assume that damage to microalgae could have repercussions throughout the entire trophic chain. Moreover, the formation of HABs and the consequent release of toxins also pose a threat to human health (Berdalet et al., 2016; Masó and Garcés, 2006). Another pivotal role performed by microalgae is bioremediation (Leong and Chang, 2020; Matsunaga et al., 1999; Priyadarshani et al., 2012) of heavy metals and anthropogenic substances released into the environment.

Although many species are affected by the impacts of climate change, showing a tendency to disappear or to be less efficient, on the bright side, there are species that are well acclimated and able to adapt to changes in the environment, enhancing their resistance to rising temperatures (Abu-Ghosh et al., 2020) and increased CO_2 levels (Guo et al., 2017).

During the last decade, scientists have attempted to develop new approaches to predict climate change (Collins, 2007), including the creation of Earth System Models, which are capable of replicating past climate conditions and are considered one of the best means of predicting future climate (Lynch, 2008). Many studies have recreated the effects of climate change by taking into account species distributions and extinctions, while also exploring the plasticity and adaptability of species physiology, combined with powerful tools for predicting ecological pathways and consequences for marine organisms (Fuller et al., 2010). However, there are limited studies regarding the use of microalgae as tools to predict global warming.

This work focuses on microalgae and how it is possible to use them as novel tools to predict and assess environmental quality. The most advanced research has seen the use of microalgae as omic tools to predict climate change (Hennon and Dyhrman, 2020), through the use of genomics, transcriptomics, proteomics, and metabolomics.

These tools are useful for predicting the trends that HABs could exhibit over the years, providing knowledge about the future occurrence of toxic and non-toxic blooms for some species of microalgae and offering a clear idea of how to prevent HABs.

7.2 THE EFFECTS OF CLIMATE CHANGE AND HARMFUL ALGAL BLOOMS (HABS)

Year by year, climate change is transforming marine waters that have already experienced progressive warming, acidification, and deoxygenation. This phenomenon will be intensified this century, and it is expected to have a direct influence on aquatic photosynthetic organisms (Song et al., 2018). These global environmental changes are happening at an exceptional speed with no precedents in the geological past (IPCC, 2014). Besides, all the transformations of environmental conditions in the sea are affecting public health, tourism, fishery, aquaculture, and ecosystems (Gobler et al., 2017).

The rise in temperature is one of the most evident consequences of climate change, as observed in the melting of glaciers and the Greenland ice cover due to this phenomenon (Teng et al., 2006). The effects will be of great magnitude: sea level rise, flooding (Frei et al., 2006), heat waves, hurricanes, droughts, and other extreme events (IPCC, 2007). Climate change is a consequence of increased atmospheric CO_2 concentration; average global sea surface temperatures are expected to increase by 1.7–1.9°C over the next century (Wigley and Raper, 2001). However, shifts in temperature will vary depending on different areas of the world's aquatic ecosystems (Beardall and Stojkovic, 2006).

Fortunately, oceans act as a buffer against climate change because of their role as carbon dioxide (CO_2) sinks (Egilsdottir et al., 2009). Under the assumption that CO_2 emissions continue without significant abatement (RCP 8.5 in the Fourth Intergovernmental Panel on Climate Change modeling convention), sea surface water temperatures (SST) are projected to increase by 2055 (Tester et al., 2020) by an average of 0.4–1.4°C in the Atlantic Ocean, 0.5–1.4°C in the Pacific Ocean, and 1.0–1.4°C in the Indian Ocean; an increase of 1.0–1.4°C in SST is predicted for the Caribbean Sea and Gulf of Mexico, and 0.8–1.4°C in the Mediterranean Sea.

However, as a consequence of CO_2 absorption, ocean chemistry is being altered, with a decrease in seawater pH (Raven et al., 2005). Ocean pH has fallen about 0.1 units since pre-industrial times (Orr et al., 2005), and it is predicted to decrease by about 0.3–0.5 units by the year 2100 and 0.5–0.7 units by the year 2300 (Caldeira and Wickett, 2003).

Changes in the chemical features of aquatic ecosystems, such as the rise in temperature and CO_2, decrease in pH, acidification, and so forth., will affect organism growth and physiological activities, but also populations indirectly (Beardall and Stojkovic, 2006), leading to dramatic changes in overall aquatic life (Abu-Ghosh et al., 2020).

The presence of CO_2 in seawater systems is fundamental for microalgae photosynthesis: an amount of 93% of CO_2 is stored in algae, vegetation, and coral (Papiol, 2018). Microalgae can be a good proxy for climate change for several reasons; since phytoplankton in nature is not exploited commercially, it is possible to draw possible links between long-term changes in population size, structure, and climate change (assuming no direct human impacts via top-down consumers, such as fish). Also, most species are short-lived, so population size is less influenced by the persistence of individuals from previous years, which leads to tight coupling between environmental change and plankton dynamics (Beardall and Stojkovic, 2006).

One example of changing population density is given by the likely increase of autochthonous dissolved organic carbon, which has long been assumed to be a major source of organic pollution to eutrophic freshwater ecosystems because it may facilitate the proliferation of cyanobacteria with a consequent bloom (Li et al., 2016).

Microalgae are affected by climate change, mainly by changes in temperature, light, OA, water stratification, and precipitation-induced nutrient inputs, but they also play a pivotal role in the regulation of Earth's climate and its cycles; for that reason, it is important to preserve them. Temperature is a key factor in the growth, abundance, and distribution of marine phytoplankton (Gobler et al., 2017; Thomas et al., 2012; Wells et al., 2015). Some microalgae populations may acclimate to changes, and some microalgae populations are able to grow when environmental conditions are favorable, reaching high densities of cells per mL in a few days to weeks. These rapid growth events, known as Harmful Algal Blooms (HABs), are usually harmful to aquatic organisms and human economic activities and, in some cases, involve the production of toxic compounds that can intoxicate people through exposure to aerosols or the consumption of aquatic organisms that contain these toxins (FAO, 2004). The presence and persistence of HABs are generally expected to be promoted by the eutrophication of coastal zones (Heisler et al., 2008); however, these blooms may be further aggravated by climate change (Dale et al., 2006; Edwards et al., 2006; Trainer et al., 2003).

The National Oceanic and Atmospheric Administration (NOAA) declared in 2019 that the past five years were the hottest on record, an observation that agrees with IPCC reports, which predict that sea surface temperatures are projected to warm by 0.4–1.4°C by the middle of this century, causing HABs of dinoflagellates such as *Gambierdiscus*, *Fukuyoa*, and *Ostreopsis*, leading to higher growth rates and greater population density (Tester et al., 2020). *Ostreopsis* is one of the most harmful algae, and its blooms have significant negative consequences for human health because they cause respiratory illness and skin irritation throughout the Mediterranean and its adjacent seas (Tester et al., 2020).

Moreover, blooms of the genera *Alexandrium* and *Gonyaulax*, studied for 20 years using field, laboratory, and sediment data, also show peaks in the numbers of *Gymnodinium catenatum*-like cysts, which correspond to intervals of climate warming during the past 2000 years (Thorsen and Dale, 1997).

The most affected group of phytoplankton responsible for HAB outbreaks comprises dinoflagellates; studies have revealed that their growth rate increases with elevated pCO_2 (Brandenburg et al., 2019). Dinoflagellates exhibit RuBisCO with low specificity toward CO_2 (Badger et al., 1998; Tortell, 2000). To overcome this disadvantage, dinoflagellates have developed highly active CO_2-concentrating mechanisms (CCMs) to ensure effective carboxylation (Ratti et al., 2007; Rost et al., 2006) and exhibit higher uptake of CO_2. Brandenburg et al. (2019) note that, for HAB species other than dinoflagellates, the number of studies is limited; therefore, more research should be conducted to better assess the responses of HAB species to elevated pCO_2.

However, increased temperature might not always have the same effect on HABs: Moss et al. (2003) observed that warming had considerably smaller effects on the phytoplankton community than nutrients did, with minor effects on chl-a and total phytoplankton biovolume. In the case of Cryptophyceae and Dinophyceae, warming did not increase their abundance (Beardall and Stojkovic, 2006).

Environmental conditions affect not only microalgal blooms but also their physiology, including their ability to produce toxins, which are dangerous for the ecosystem and human health (Anderson et al., 2012). Generally, the higher the temperature, the greater the production of toxins, and seafood is also likely to be affected by climate change (Miraglia et al., 2009).

The production of volatile toxic compounds is also due to metabolic stress (Vardi et al., 2006) under different environmental conditions. These compounds can affect the offspring of zooplankton and other aquatic organisms, as well as their reproductive success (Barreiro et al., 2011; Ragg et al., 2010). It is not in all areas that there is an increase in nutrient input. Eutrophication is related to the presence and persistence of toxin- and toxin-producing species (Moore et al., 2009). Still, there is agreement that due to the increase in average temperature, the distribution of tropical harmful microalgae species will increase by colonizing ecosystems in higher latitudes; but even with the prediction that with climate change the number of HABs will increase, their likely persistence and toxicity is difficult to predict (Papiol, 2018) because species responses change. Among the toxic compounds produced by HABs, we can mention the production of domoic acid (DA) in *Pseudo-nitzschia* spp., which has been shown to increase significantly under high CO_2 conditions (Sun et al., 2011; Tatters et al., 2013).

The response of toxic substances to CO_2 changes is different even among the same species but different strains: the increase of CO_2 leads to a decrease in paralytic shellfish poisoning (PSP) toxin content in *Alexandrium catenella* (Van De Waal et al., 2014), while PSP toxin contents in the *Alexandrium catenella* strain increased with elevated pCO_2 (Tatters et al., 2013).

Not only changes in CO_2 concentration may influence the release of toxic compounds from microalgae, but also temperature changes. For instance, the toxicity of *Protoceratium reticulatum* attributed to yessotoxins has been shown to increase with higher temperatures (Guerrini et al., 2007; Paz et al., 2006), whereas the dinoflagellate *Karenia brevis* has demonstrated higher toxicity at lower temperatures (Brandenburg et al., 2019).

7.3 FACTORS THAT AFFECT MICROALGAE CAUSED BY CLIMATE CHANGE

The factors that affect microalgae due to climate change are temperature rise, OA, pH change, UV-B radiation, and salinity change. All of them stem from the excess amount of CO_2 in the atmosphere that affects seawater and organisms, including microalgae.

7.3.1 INCREASE OF CO_2

Increase of CO_2 in seawater also leads to changes in the photosynthesis rate. Microalgae use ribulose bisphosphate carboxylase-oxygenase (RuBisCO) as the primary CO_2-fixing enzyme in carbon assimilation. Since RuBisCO has a low affinity for CO_2, the uptake of CO_2 is limited under present-day CO_2 levels (Beardall and Raven, 2004). Organisms that uptake low levels of CO_2 have evolved CCMs that actively transport and accumulate inorganic carbon and hence increase the concentration of CO_2 at the active site of RuBisCO. CCMs are regulated by environmental factors such as light intensity and spectral quality, nutrient status, and other factors that affect the availability of CO_2 (Beardall and Raven, 2004). Without CCMs, some organisms would be unable to carry out net photosynthesis under air-equilibrium CO_2 and O_2 concentrations (Raven et al., 2000). Moreover, temperature also influences the kinetics of RuBisCO: increasing temperature results in an increase in the substrate-saturated specific reaction rate and a decrease in the affinity for CO_2 (Beardall and Raven, 2004).

The responses of photosynthesis to inorganic carbon concentration are likely to differ among organisms in response to current and future changes in atmospheric CO_2. For instance, photosynthesis by *Phaeodactylum tricornutum* increases substantially as external CO_2 rises, whereas photosynthesis by *Thalassiosira weissflogii* shows a much smaller increase when cells are grown above 180 ppm CO_2 (Beardall and Raven, 2004). This indicates that differences in photosynthetic affinity for CO_2 result in differences in growth rate responses to CO_2 (Beardall et al., 1998; Raven et al., 1993).

Organisms such as the picoplanktonic Chlorophyta alga *Stichococcus bacillaris* (formerly *Stichococcus minor*) (Munoz and Merrett, 1989), the Chlorophyta lichen symbiont *Coccomyxa* (Palmqvist, 2000), and some strains of the nanoplanktonic Apatophyta alga *Emiliania huxleyi* (Badger et al., 1998; Nimer and Merrett, 1992) show low affinity for CO_2 in photosynthesis and do not begin to exhibit saturation of photosynthesis until dissolved CO_2 approaches 25 μM (Raven and Johnston, 1991).

Photosynthesis by the marine diatom *Thalassiosira pseudonana* shows decreased affinity for inorganic carbon as the inorganic carbon concentration provided during growth increases (Fielding et al., 1998).

The coccolithophorids *Emiliania huxleyi* and *Gephyrocapsa oceanica* respond differently to increases in atmospheric CO_2: increases in particulate organic carbon (POC) production in response to CO_2 enrichment were more evident in *G. oceanica,* while *E. huxleyi* was less affected by the decrease in calcification following CO_2 enrichment and the consequent decrease in seawater pH compared to *G. oceanica.* Nevertheless, elevated CO_2 led both species to decreased calcification, lower ratios of calcite to organic carbon, and more frequent coccolith malformations (Beardall and Raven, 2004).

Moreover, increased CO_2 on the photosynthesis and growth of microalgae could lead to a reduction in the uptake of other resources fundamental for growth, such as light, N, Fe, and Zn (Beardall and Raven, 2004).

7.3.2 TEMPERATURE RISE

Relying on the change in temperature due to climate change, classes of microalgae also respond differently to temperature changes depending on their different ranges of tolerance and physiological responses (Reynolds, 1989). Goldman and Ryther (1976) showed that in mixed cultures of marine microalgae, the diatom *Skeletonema costatum* dominates at low temperatures (<10°C), whereas *Phaeodactylum tricornutum, Nitzschia hassall*, and *Entomoneis alata* become more abundant as temperature increases (25–30°C).

Changes in temperature due to climate change may also lead to indirect effects like the deposition of aeolian iron in the oceans (Duce and Tindale, 1991; Murray et al., 1995), increasing the availability of iron, or the alteration of the scheme of upwelling events, which could change the occurrence of associated phytoplankton blooms as happened in the tropical Atlantic during the last deglaciation. Those oscillations in oceanic production were related to climatic changes at high latitudes and were caused by changes in the strength or duration of upwelling events, associated with alterations in trade-wind strength, which are in turn influenced by the thermohaline circulation and sea surface temperatures (Beardall and Raven, 2004).

7.3.3 UV-B RADIATION

Among the threats of climate change, solar UV-B radiation (UVBR) is also known to have a range of deleterious effects on phytoplankton, even in conditions not enhanced by climate change. The primary production by phytoplankton in Antarctic surface waters exposed to elevated UV under the ozone hole can be inhibited by 30% (Prézelin et al., 1993), and photosynthesis by temperate phytoplankton populations in surface waters offshore of Patagonia was inhibited by ~37% (Helbling et al., 2001).

Nevertheless, there are species with high resistance to UV, like benthic microalgae exposed to high UV radiation on intertidal mudflats (Peletier et al., 1996; Roux et al., 2002). Other effects of UV on microalgae include inhibitory effects on nutrient uptake (Behrenfeld et al., 1995; Hessen et al., 1995), damage to DNA (Karentz et al., 1991), uptake inhibition, and acquisition of nutrients, such as nitrate or ammonium (Behrenfeld et al., 1995; Döhler, 1985; Fauchol et al., 2000) and phosphorus (Hessen et al., 1995).

In the Chlorophyta *Dunaliella tertiolecta*, inorganic carbon assimilation is inhibited by UVBR, but the activity of the CCM based on inorganic carbon transport is relatively less affected. Thus, the affinity of cells for inorganic carbon in photosynthesis is unimpaired or even greater after a period of UVBR exposure (Beardall et al., 2002).

Another consequence of increased levels of anthropogenic CO_2 is the rise of OA. It is known to affect the physiology and growth of marine phytoplankton, especially in polar regions (Tan et al., 2019).

7.3.4 OCEAN ACIDIFICATION AND pH

The increase of anthropogenic CO_2 in the atmosphere leads to OA, an increase in the absorbed CO_2 in the ocean due to the carbonate chemistry, reducing pH in water. The IPCC (2014) forecasts up to a 0.4 decrease in pH. OA can affect microalgae metabolism: ocean phytoplankton seems more resilient and maintains its growth rate compared to coastal phytoplankton, which seems more adaptable (Berge et al., 2010). By groups, the calcification of *Coccolithophorids' coccoliths* (Riebesell et al., 2000), as well as the formation of diatom silica frustules (Archer et al., 2000), are impaired by these changes in pH, and eventually affect their fitness and success (Papiol, 2018). Calcified organisms are very significant components of pelagic ecosystems, and any changes in the functioning of these calcified components in response to surface OA would have large implications for the functioning of the ecosystem (Raven et al., 2005). Over the past decade, OA has become a major concern for its significant ecological effects on marine organisms and potential perturbations of the carbonate system (Fabry et al., 2008; Guinotte and Fabry, 2008). A variety of responses of living individuals and their environment to OA have been studied across a global range, such as effects on organisms (Kleypas et al., 2006), changes in toxicity (Roberts et al., 2013), and behavior of natural elements (Ivanina and Sokolova, 2015).

Data from the IPCC report of 2014 indicate an increase in atmospheric CO_2 levels from the present-day 410 ppm to 1000 ppm by the end of the century unless significant mitigation measures are taken (Tan et al., 2019).

Since 1800, oceans have absorbed 26% of annual anthropogenic CO_2 (Sabine et al., 2004). CO_2 dissolves into seawater and is then converted to bicarbonate (HCO_3^-) and carbonate (CO_3^{2-}) ions, as well as hydrogen ions (H^+), the latter contributing to the acidification of the oceans, with predictions that ocean pH will drop by approximately another 0.3–0.32 units by the year 2100 (IPCC, 2014; Xia et al., 2018).

Covering 20% of the global oceans (Deppeler et al., 2018), the polar Southern Ocean (SO) is the largest carbon sink of all global oceans, taking up 43% of the CO_2 produced by human activities (Frölicher et al., 2015). By the end of the century, carbon uptake of the SO is expected to increase by a further 20%, with biological uptake predicted to double (Hauck and Völker, 2015).

With the increasing incidence of OA, the environmental factors, such as pH, salinity, illumination, and temperature, show changes (Chen et al., 2018). For phytoplankton, high extracellular pH may cause important alterations in the membrane transport processes and metabolic functions involved in internal pH regulation (Hansen, 2002; Smith and Raven, 1979). Therefore, variation in pH can reduce or increase the toxicity and availability of many substances, in particular weak acids and bases (Chen et al., 2018) in the seawater.

Moreover, changes in carbon chemistry have a severe impact on calcifying marine organisms, such as corals and coralline algae (Kuffner et al., 2008). The effects of pH on the growth of marine phytoplankton have been well documented. It has been suggested that some species, such as *Ceratium tripos* and *Tripos furca* (formerly *Ceratium furca*) (phylum Miozoa), stop growing at a pH above 8.3–8.4, whereas others are able to grow at a pH close to 10, such as *Phaeodactylum tricornutum* (phylum Bacillariophyta) and *Rhodomonas salina* (phylum Cryptophyta), based on data from 35 species of marine phytoplankton (Hansen, 2002).

In the study by Chen et al. (2018), the growth rate of *Chlorella salina* (phylum Chlorophyta) increased with increasing pH values, and the optimum pH was 8.0 under the present culture conditions. These results corresponded to the normal variation of pH in seawater at 35 salinity (Knutzen, 1981). The responses to OA vary from species to species (Johnson et al., 2013). There are several bloom-forming genera, such as the oceanic *Trichodesmium*, *Synechococcus*, and *Crocosphaera*, which increase their rates of photosynthesis, nitrogen fixation, and growth in response to CO_2 enrichment, whereas *Prochlorococcus* (phylum Cyanobacteria) has shown only a limited response (Fu et al., 2007).

7.3.5 CHANGE IN SALINITY

Salinity change in water is one of the threats and concerns related to climate change. Salinity is expected to change as a result of this global warming phenomenon, being reduced or increased depending on the region's latitude (Bautista-Chamizo et al., 2018). Actually, an increase of 0.2 is verified in the Caribbean Sea and Gulf of Mexico, and about 0.2–0.3 in most of the temperate and tropical Atlantic Ocean is predicted. Increases in the Mediterranean Sea are expected to be more variable, ranging from no change in the eastern basin to an increase of up to 0.4 in the western basin (Durack and Wijffels, 2010).

This factor is related to changes in temperature and pH, and models show a general decrease in salinity of 0.1–0.3 in the central and west-central Pacific Ocean, the eastern and central Indian Ocean warm pool regions, and the Baltic Sea by 2055 (Tester et al., 2020).

Only a few research studies are focused on the effects of CO_2 enrichment when combined with changes in temperature or salinity (Ko et al., 2014; Matozzo et al., 2012; Pereira et al., 2016), but there is still a lack of research about the potential effects of these three combined environmental factors on microalgae (Bautista-Chamizo et al., 2018).

Bautista-Chamizo et al. (2018) worked to provide an integrated understanding of the microalga *P. tricornutum* response, considering the effects of three variables (temperature, salinity, and pH) affected by global warming. Three pH values (6, 7.4, and a nonacidified control, pH = 8), two temperature values (23 and 28°C), and two salinity values (34 and 40 psu) were selected to investigate the effects on key physiological parameters of *P. tricornutum*. After 48 hours of exposure, esterase activity analysis revealed that cell metabolic activity appeared to be exacerbated by high temperature and salinity, and low cell density was detected at pH 6 (28°C), where the number of cells was reduced by approximately threefold under salinities of 34 and 40 psu. Their study revealed the negative effect provoked by CO_2 enrichment in seawater on *P. tricornutum*. Nevertheless, the effects caused by this phenomenon, when combined with other environmental stressors, yield different responses among species, and further studies are required to verify the synergistic effects of the three parameters together (Bautista-Chamizo et al., 2018).

The effects of increased temperature, CO_2 availability, and salinity change were also studied on 16 genetically different isolates of the diatom *Skeletonema marinoi* from the Adriatic Sea and the Skagerrak (North Sea), and on eight strains of the dinoflagellate *Alexandrium ostenfeldii* from the Baltic Sea (Kremp et al., 2012).

In both species, individual strains were affected differently by increased temperature and pCO_2. A high degree of response variability was detected among individual strains of *S. marinoi* and *A. ostenfeldii*. The *S. marinoi* strains from the Adriatic buffered the overall effects of increased CO_2 supply and temperature. The relatively uniform response of *S. marinoi*, with many strains exhibiting increased growth rates at higher temperatures, resulted in an overall positive effect of temperature on growth in this population. Depending on the strain, temperature and pCO_2 caused positive, negative, or no effects. This emphasizes that responses observed in single-strain experiments may not be representative and that predictions of species behavior under future climatic conditions need to be interpreted with caution (Kremp et al., 2012).

7.4 RESPONSES OF MICROALGAE TO CLIMATE CHANGE FACTORS: THREATS AND ACCLIMATION

There is evidence of the impact on marine algae by climate change (Brodie et al., 2009). For instance, several species have moved north of their own accord instead of having been intentionally or accidentally introduced (Brodie et al., 2009; Sauvageau, 1916).

For example, it has been reported that the genetic composition of the endosymbiotic dinoflagellates (*Symbiodinium*) associated with coral colonies varies in response to environmental oscillation (Baker et al., 2004; Thornhill et al., 2006), such as increases in temperature related to climate change. Sassenhagen et al. (2015) observed that microalgae could grow under a large range of environmental conditions because of high phenotypic plasticity.

Each species needs different approaches if threatened. The response is conditioned by the threat. If the threat is global warming/climate change, it could be relevant to consider where species might move towards and to maintain those environments in good condition; therefore, all habitats need to be looked after (including buffer zones) (Brodie et al., 2009).

It is not always possible to protect a habitat, and thus other means of protecting threatened and endangered species are required. One strong means is *ex situ* cultivation (maintaining the species outside its natural habitat), and this approach is specifically applicable to microalgae (Brodie et al., 2009). For example, if an organism is limited to salinities lower than 40 psu, then it cannot occur in hypersaline environments. In similar ways, pH, nutrient concentrations, alkalinity, metal concentration, and so on provide upper and lower limits of survival. This is called Shelford's Law of Tolerance (Shelford, 2013), and one cannot expect organisms to exist where they are not able to survive (Brodie et al., 2009).

7.4.1 LIGHT AND TEMPERATURE

Due to global warming and changing environmental conditions, such as light intensity and water temperature in the coming years, climate change is expected to have a direct influence on photosynthetic aquatic organisms (Song et al., 2018), which will affect their growth and physiological activities, leading to noticeable changes in overall aquatic life (Abu-Ghosh et al., 2020). The ocean has absorbed more than 90% of the excess energy in the climate system, resulting in warming of the upper 75 m by 0.11°C–1°C between 1971 and 2010 (Rhein et al., 2013). End-of-century model projections for average sea surface temperatures forecast increases ranging from (mean ± SD) 0.71 ± 0.45°C (Representative Concentration Pathway [RCP] 4.5) to 2.73 ± 0.72°C (RCP 8.5) (Howes et al., 2015). Several regional "hot spots" were projected under RCP 8.5 in the Arctic, tropics, and North Pacific, with increases exceeding 4°C (Bopp et al., 2013; Howes et al., 2015). As anthropogenic climate change raises average temperatures, marine heat waves are expected to increase (Schmidt and Boyd, 2016).

Every species of microalgae has an optimal combination of temperature and light, resulting in maximum growth rates with less alteration of physiological or biochemical parameters at the beginning of their acclimatization (Singh and Singh, 2015). The majority of microalgae have the capacity to carry out photosynthesis and growth under temperatures between 15°C and 30°C, with optimal conditions, in most cases, between 20°C and 25°C (Raven and Geider, 1988).

Microalgae experience light and temperature differences primarily between the upper and lower two limits when moving from the surface to deeper water levels, thus facing physical parameters below or above the physiological thresholds of most algae species commercialized (Mata et al., 2010).

Abu-Ghosh et al. (2020) studied the effect of acclimation to elevated temperatures on photosynthesis and photosynthesis-related activities using the aquatic thermotolerant Chlorophyta *Micractinium* sp., which has the capacity to acclimate to a wide temperature range of 10–50°C (Sonmez et al., 2016), combined with photoacclimation to limiting and saturating light conditions (Adar et al., 2016). *Micractinium* sp. cells exhibited the highest growth and net oxygen production rates at the optimal temperature of 25°C, independent of the light acclimation conditions. The results of Abu-Ghosh et al. (2020) indicate that the upper thermal threshold for *Micractinium* sp. photosynthesis and growth ranges between 35°C and 40°C. This microalga was more susceptible to temperature change when acclimated to limiting light levels, whereas it exhibited stable photosynthetic efficiency and effective nonphotochemical quenching (NPQ) under saturating light. These results show that acclimation of thermotolerant microalgae to saturating light enhances the thermal tolerance of PSII. This characteristic results from enhanced heat stability of PSII photochemistry and oxygen evolution. Some studies have already shown an increase in specific growth rates with increasing light supply (Abu-Ghosh et al., 2015; Gonçalves et al., 2016; Ota et al., 2015).

Nevertheless, in the light–temperature matrix, there was a strong interaction suggesting a complex synergy among these environmental factors determining the growth of *Micractinium* sp. (Abu-Ghosh et al., 2020). This response to elevated temperatures is similar to the differential regulation observed in *Trachydiscus minutus* (class Eustigmatophyceae) (Gigova et al., 2012) and in *Dunaliella viridis* (phylum Chlorophyta) (Srirangan et al., 2015) when exposed to moderate light of 130–260 μmol photons m^{-2} s^{-1} at 25°C (Abu-Ghosh et al., 2020). Thus, the results indicate that the relationship between photosynthesis and irradiance parameters and the acclimating light conditions is highly dependent on growth temperature (Abu-Ghosh et al., 2020).

The photophysiology and pigment composition of *Micractinium* sp. acclimated to elevated temperatures were significantly influenced by changes in light intensity, which led to differences in photosynthetic activities and, eventually, growth. These results indicate that thermotolerant microalgae have developed mechanisms to increase light absorption and utilization efficiency at elevated temperatures (Abu-Ghosh et al., 2020).

Photo-acclimation involves an optimization of cellular pigment composition based on available light flux, increasing light-harvesting pigments under faint light, and inversely their decrease in response to high light, concomitant with an increase in photoprotective pigments (MacIntyre et al., 2002; Magdaong and Blankenship, 2018).

There is an incomplete understanding of the physiological mechanisms involved in the acclimatization of photosynthesis to temperature. Light is another important factor to consider regarding the response of photosynthesis to temperature (Abu-Ghosh et al., 2020). There is increasing recognition that changes in light and temperature factors are connected, and their effect on photosynthesis should be investigated when combined, rather than primarily showing growth rate kinetics (Singh and Singh, 2015).

Algae and other plants frequently develop physiological adaptations when confronted with modifications in their natural environment (Michel et al., 1989). This is true for temperature, which is one of the principal factors that control metabolic rates and photosynthesis (Michel et al., 1989). For instance, plants growing in cold environments normally show lower optimal, minimal, and maximal temperatures for photosynthesis than plants from warmer climates (Berry and Bjorkman, 1980; Li, 1985).

To assess the effects of rapid changes in temperature on the photosynthetic activities of Arctic sea-ice microalgae, photosynthetic acclimatization to temperature may be of great ecological significance in extending the growth season of ice algae (Michel et al., 1989). Salinity at the ice–water interface of polar seas is normally high; thus, microalgae grow under conditions of very low temperature (down to < –1.5°C) (Meguro et al., 1967). It has been suggested that ice algae are closely adapted to these low environmental temperatures and are therefore psychrophilic (Bunt, 1964; El-Sayed and Taguchi, 1981; Van Baalen and O'Donnell, 1983).

Palmisano et al. (1987) reported, for Antarctic sea-ice microalgae, an increase in both maximum photosynthetic rate and photosynthetic efficiency with increasing temperature; their optima occurred at 8°C and 2°C, respectively. Other studies of Antarctic phytoplankton observed similar increases in photosynthetic rates at both saturating and limiting light levels (Tilzer et al., 1986), as well as increases in assimilation numbers up to temperatures ranging from 4.5 to 7°C (Neori and Holm-Hansen, 1982) and up to 12°C (Jacques, 1983). These results support the idea that the optimal temperature for photosynthesis is normally higher than the environmental temperature and that physiological processes in nature usually occur at suboptimal temperatures (Li, 1980). Li et al. (1984) observed that Arctic phytoplankton photosynthesis is likely limited by enzyme reactions at suboptimal temperatures, in agreement with other studies (Berry and Bjorkman, 1980; Berry and Raison, 1981; Li et al., 1984; Nielsen and Hansen, 1959).

As the growth season progressed, ice algae became acclimatized to modifications in temperature that occurred in the natural environment (Michel et al., 1989). *N. frigida* may be particularly well suited to cope with the environmental conditions at the ice-water interface (Michel et al., 1989; Rochet et al., 1985). Temperature acclimatization could be of high ecological relevance in extending the growth season of ice algae by several weeks (Michel et al., 1989).

7.4.2 IRRADIANCE AND PHOTOSYNTHETICALLY ACTIVE RADIATION (PAR)

Due to stratospheric ozone depletion, UVBR levels are enhanced during springtime in polar regions, which may cause an integrated loss of carbon fixation in Antarctic waters ranging from <0.15% to 12% (Helbling et al., 1994; Holm-Hansen et al., 1993; Smith et al., 1992). Acclimation to ultraviolet-A radiation (UVAR: 315–400 nm) or to high irradiances of photosynthetically active radiation (PAR: 400–700 nm) may induce the production of mycosporine-like amino acids (MAAs), which protect cells against incoming short wavelengths (Banaszak, 2003; Neale et al., 1998b). In addition, when cells are acclimated to high *in situ* irradiances, repair mechanisms, such as enzymes that correct UVBR-induced DNA damage, may be more active (as occurs in higher plants) (Takahashi et al., 2002; Waterworth, 2002). It is unknown how photooxidative and antioxidative mechanisms affect UVBR sensitivity. Microalgae, like the majority of organisms, develop or induce mechanisms that provide protection against excess solar irradiance (Banaszak, 2003).

A study carried out by Buma et al. (2006) examined the effect of PAR acclimation of four common Antarctic marine microalgae (the dinoflagellate *Polarella glacialis*, the haptophyte *Phaeocystis antarctica*, the diatom *Chaetoceros dichaeta*, and the Chlorophyta *Pyramimonas gelidicola*) on their sensitivity to UVBR-induced DNA damage. The results show that PAR acclimation affected the biovolumes of the dinoflagellate *Polarella glacialis* and the haptophyte *Phaeocystis antarctica*. PAR acclimation strongly affected the pigment content and/or ratio in all species.

Other studies reported increased UVBR tolerance (PSII efficiency and viability) in high-PAR-acclimated diatoms (Van de Poll et al., 2005). In addition, lower UVBR vulnerability was also observed when cells had been acclimated to high PAR first, in comparison with low-light-acclimated cultures (Litchman and Neale, 2005; Neale et al., 1998). It has also been suggested for microalgae (Davidson, 2006) that there is strong habitat-related vulnerability to UVBR (Van de Poll et al., 2001). It is probable that organisms adapted to high mean irradiance habitats are well equipped to overcome, simultaneously, high PAR, UVAR, and UVBR stress (Buma et al., 2006).

7.4.3 OCEAN ACIDIFICATION AND CO_2 INCREASE

The concentration of hydrogen ions in the Earth's oceans is rising due to atmospheric increases in anthropogenic CO_2 (Reid et al., 2019). The resultant decrease in pH is called ocean acidification (Pörtner, 2008). Coastal and estuarine pH can also vary because of natural variability, eutrophication, net heterotrophy (Borges and Gypens, 2010; Kemp et al., 2005), acid-forming compound deposition (Doney et al., 2012), regional changes in land use (Green et al., 2009), and watershed inputs (Dove and Sammut, 2007; Salisbury et al., 2008). The increasing partial pressure of carbon dioxide (pCO_2) causes a decrease in the seawater saturation states of aragonite and calcite, the most commonly biomineralized forms of calcium carbonate (Miller et al., 2009) employed for the shell formation of some marine organisms.

Warmer, more layered oceans are expected to modify phytoplankton communities to a smaller size with reduced adaptive capacity (Acevedo-Trejos et al., 2014) and shift ranges of cold-water species toward the pole (Hallegraeff, 2010). A major concentration of dissolved CO_2 due to climate warming is also expected to reduce the omega-3 fatty acid (FA) content in phytoplankton, the main source of omega-3 FAs in aquatic systems (Hixson and Arts, 2016).

To mitigate stress caused by a high CO_2 environment, microalgae can develop adaptive mechanisms (Baba and Shiraiwa, 2012; Brueggeman et al., 2012; Cuellar-Bermudez et al., 2015; Giordano et al., 2005; Jia et al., 2015; Solovchenko and Khozin-Goldberg, 2013; Solovchenko et al., 2015; Wang et al., 2011).

The observed responses to high CO_2 concentrations include regulations of CO_2 concentrating mechanisms (CCMs), rearrangements and transitions of membrane proteins, inhibitions of photorespiration, and triggering gametogenesis (Guo et al., 2017). In *Chlamydomonas reinhardtii*, the CCMs would be inhibited by the bicarbonate transporters and carbonic anhydrases (CAs) in the elevated concentration of CO_2 (Guo et al., 2017). Nevertheless, there is not much knowledge about the responses of the tolerant microalgae (Guo et al., 2017).

The majority of molecular acclimatization studies reported are based on *C. reinhardtii* photosynthesis activity, which is low at high CO_2 concentrations. Few conclusions can be drawn on the mechanisms for a tolerant species (Baba and Shiraiwa, 2012; Boyle and Morgan, 2009; Brueggeman et al., 2012; Chang et al., 2011; Dishisha et al., 2014; Giordano et al., 2005; Renberg et al., 2010; Solovchenko and Khozin-Goldberg, 2013; Swarnalatha et al., 2015; Tripp et al., 2010; Wang et al., 2011; Winck et al., 2016).

The continuous transfers are an adaptive procedure which, integrated with the gas bag system, possesses the potential to develop a tolerant species. Under the stress of the elevated concentration of CO_2, the microalgae transfer the genetic information for acclimation to the offspring. Being genetic information, it is highlighted and maintained during the continuous transfers (Collins and Bell, 2004).

Responses to increased CO_2 are different among phytoplankton functional groups identified in calcifiers, silicifiers, nitrogen fixers, and other greens; each group could show variation in plasticity and evolution (Collins et al., 2013; Dutkiewicz et al., 2015; Pardew et al., 2018; Schaum et al., 2013). This means that variation may also bring a positive response in terms of nutrient enrichment due to an increase in carbon availability, translating into an increase in the population growth rate (Dutkiewicz et al., 2015; Follows et al., 2007).

Species able to adapt and increase in population under difficult conditions are called "climate change winners" (Dutkiewicz et al., 2013; Lindberg and Collins, 2020). Our knowledge of evolutionary processes is based principally on experiments performed in stressful, starvation, or toxic environments (Van den Bergh et al., 2018).

7.4.4 ADAPTABILITY TO CHANGE IN pH

Majority of microalgae cannot survive at low pH (<6.0) (Perez-Garcia et al., 2011). Jiang et al. (2012) observed that microalgae non-adapted to acidic conditions could not survive at pH 3.0, showing that acid adaptation is imminent for survival. In addition, microalgae grown at pH 4.0 showed a drastic decline in biomass, suggesting the significant toxicity of pH (Khalil et al., 2010). Besides, pH 4.5 inhibited 50% of growth in acid-tolerant microalgae (Nalewajko et al., 1997).

El-Ansari and Colman (2015) observed that acid-tolerant microalgae could not grow at pH 3.0 due to a decrease in intracellular pH. Therefore, acid-tolerant microalgae are also sensitive to low pH, exhibiting that only acidophiles have the capacity to grow under such extreme conditions due to gene inheritance through evolutionary response (Hirooka et al., 2017). This is due to distinct environmental pressures like acidic conditions; only limited strains of non-acidophilic microalgae could phenotypically adjust to thrive and grow (Abinandan et al., 2018).

Evidence of the impact of climate change, OA, and introduced species on native flora is usually anecdotal for marine macroalgae as well as marine microalgae, underscoring the need for long-term monitoring and scientific studies to determine the changes in abundance and distribution. As ways to move forward for algae conservation and public awareness, the use of species and site designations is studied, including biodiversity action plans, major plant areas, and key biodiversity areas (Brodie et al., 2009).

7.5 THE DUAL ROLE OF PHYTOPLANKTON IN ECOSYSTEMS

Coastal, estuarine, and lagoon ecosystems are being subjected to several pressures of anthropogenic and natural origin. This endangers water resources, which are already scarce, and directly affects the economy, ecosystems, and public health (Alda-Vidal et al., 2020). For this reason, the

2030 Agenda and Sustainable Development Goals comprise water resource management and improvement at a global level (United Nations [UN], 2015). In addition, solid waste such as plastics and personal care products (Spence et al., 2016), as well as emerging contaminants dissolved in water (e.g., microplastics, organic and inorganic substances, pharmaceuticals, and metals), can be detected in water or in organisms that inhabit these ecosystems due to recent technological advances (Chowdhury et al., 2016; Eerkes-Medrano et al., 2019; Emparan et al., 2018; Menció et al., 2016; Schaum, 2018). Phytoplankton, which form the basis of food webs, hold a dual role in aquatic ecosystems, being considered both environmental quality bioindicators and bioremediation agents. Community composition differs according to habitat characteristics. While Cyanobacteria are abundant in eutrophic waters (e.g., lakes), diatoms (phylum Ochrophyta) are dominant in oligotrophic ecosystems (Wetzel, 2001). Microalgae and cyanobacteria are part of phytoplankton, and their biomass fluctuates seasonally according to variations in biotic and abiotic parameters (Le Faucheur et al., 2014).

For instance, the presence of metals, pesticides, or other pollutants in water bodies has harmful consequences at all levels of biological organization (Balzano et al., 2020). As phytoplankton is at the base of the food chain, biomagnification of the pollutant occurs throughout the trophic chain, affecting everything from the cell to the community (Wu et al., 2020). For example, the metal mercury (Hg) is accumulated by phytoplankton and can also cause physiological changes in their cells (Aoki et al., 2008; Le Faucheur et al., 2014). Thus, grazers and other organisms that feed on phytoplankton will thereby accumulate Hg (Pomerleau et al., 2016).

Another toxic metal, thallium (Tl), can be assimilated into the phytoplankton cytosol and therefore become available to higher trophic levels. Among the species tested, Zhang and Rickaby (2020) found that the cyanobacterium *Synechococcus* sp. (PCC6803) and the haptophytes *Isochrysis galbana* and *Pavlova granifera* are able to grow normally at Tl concentrations up to 1 mg/L (Zhang and Rickaby, 2020). Several phytoplanktonic microorganisms are bioindicators of good or poor aquatic ecosystem quality (Filimonova et al., 2018; Pham, 2020; Tagliaferro et al., 2018). Due to phytoplankton resilience and contaminant tolerance, it is possible to better understand pollution events (Arce et al., 2006; Molina-Santos et al., 2018).

In fact, phytoplankton is able to remove several types of pollutants from aquatic ecosystems, such as inorganic load (Wang et al., 2017), metals (Wollmann et al., 2019), pharmaceuticals (Gentili and Fick, 2017; Leng et al., 2020), radioactive minerals (Varjani et al., 2018), or pathogenic organisms (Marchello et al., 2015). Phytoplankton can metabolize or compartmentalize toxic compounds in specific organelles, which, from another perspective, makes them suitable as bioremediation or clean-up agents (Kumar et al., 2015). For instance, several microalgae species have already proven to be an efficient biological way to treat effluent, wastewater, or polluted sites. A study performed by Singh et al. (2020) showed that the microalgae *Chlorella* sp., *Chlorococcum* sp., and *Neochloris* sp. are able to remove organic pollutants while enhancing fatty acid production through cultivation in river water contaminated with pharmaceuticals.

The large-scale application of pesticides leads to a high content that enters aquatic ecosystems through lixiviation, and they are also pollutants that, through biomagnification, reach high trophic levels with harmful impacts on human health (Boudh and Singh, 2018). In this context, the green microalga *Chlorella vulgaris* is considered a resilient species that can effectively remove a wide range of pesticides (87%–96% removal), such as atrazine, molinate, simazine, propanil, and dimethoate (Hussein et al., 2017). Nevertheless, it is necessary to consider that different microalgal species exhibit different contaminant removal rates (Jin et al., 2012; Kurade et al., 2016). Foremost, it is important to screen microalgae to characterize the microalgal community at the contaminated site both molecularly and morphologically (Wu et al., 2013). Subsequently, assessment of the metabolic mechanisms involved in pollutant removal is pivotal to achieving successful contaminant removal (Nie et al., 2020). Microalgae can clean up aquatic ecosystems through bioadsorption, a process by which they remove pesticides, organic substances, or aromatic compounds (Ardal, 2014; Hussein et al., 2017). This metabolic mechanism is passive (e.g., electrostatic interaction, surface complexation, ion exchange, absorption, or precipitation) and is influenced not only by microalgal cell wall composition but also by other abiotic parameters, such as pH, temperature, redox potential, light, and other electron acceptors (Abdel-Ghani and El-Chaghaby, 2014; Bilal et al., 2018; Fomina and Gadd, 2014; Priyadarshani et al., 2012).

The question that arises is: what happens to the pollutant when adsorbed by microalgal cells? Further research reveals that they are capable of bioaccumulating and biodegrading these pollutants in order to incorporate them into their metabolism (Matamoros et al., 2016; Pérez-Legaspi et al., 2016; Xu and Huang, 2017). The performance of three filamentous cyanobacteria in the removal of an organophosphorus pesticide was evaluated, namely *Nostoc oryzae* (previously known as *Anabaena oryzae*), *Desmonostoc muscorum* (formerly known as *Nostoc muscorum*), and *Spirulina* (*Arthrospira*) *platensis*. Researchers concluded that despite the low growth rate, these cyanobacteria were able to effectively remove this contaminant from the water and incorporate it into their metabolism (Ibrahim et al., 2014).

Tetradesmus obliquus (formerly known as *Scenedesmus obliquus*) was also tested for its ability to enantioselectivity biodegrade a toxic compound (triadimefon) into its metabolite triadimenol (Xu and Huang, 2017).

In this process of biotransformation, several metabolic components are involved, namely the enzymes esterase, transferase, and cytochrome P450 (Devonshire and Field, 1991). While for biodegradation, enzymes such as phosphatase, phosphodiesterases, oxygenase, and oxidoreductase are involved (Kanekar, 2004; Ortiz-Hernández, 2013). In this context, phytoplankton communities in aquatic ecosystems assume their importance as natural bioremediatory agents of the environment.

7.6 EFFECTS OF CLIMATE CHANGE ON FOOD WEBS AND OMIC TOOLS TO PREDICT CHANGES IN THE MARINE COMMUNITY

Climate change is a threat not only to species directly involved, but there are also repercussions for species linked to the trophic chain. Indeed, when phytoplankton decreases, there are consequences for all the trophic levels. Scientists agree on the decline of global primary production in response to climate change, largely due to the reduction of the supply of nutrients to the upper ocean layer. Evidence from data suggests a potential amplification of this response throughout the trophic food web, with more dramatic responses at higher trophic levels (Kwiatkowski et al., 2019). A probable scenario foresees that phytoplankton biomass would decline by 6% and zooplankton biomass by 11% over the 21st century (Chust et al., 2014), representing negative global trophic amplification between phytoplankton and zooplankton (Kwiatkowski et al., 2019).

One of the limitations of studies on global trophic interactions under climate change is that they are typically dependent on ocean biogeochemical (OBGC) models, which assume phytoplankton stoichiometry is fixed at constant Redfield values with respect to either C:N:P (Chust et al., 2014) or C:N (Stock et al., 2014). Unfortunately, fixed stoichiometry models could miscount the measure of net primary production declines in response to increased nutrient limitation (Flynn, 2010; Kwiatkowski et al., 2018; Teng et al., 2014). Still, these models do not account for the effect of changing phytoplankton stoichiometry on the food quality available to higher trophic levels, such as zooplankton (Mitra and Flynn, 2005). The most evident effect caused by climate change on the marine trophic web is the spreading of HABs, which affect directly and indirectly all marine organisms.

Through the formation of HABs, phytoplankton affects not only ecosystems but also fishery, tourism, and human health by the production of toxins accumulated in the food chain that lead to the death of fish, seabirds, and marine mammals (Anderson et al., 2002; Hallegraeff, 1993).

A common harmful consequence of toxin release is the accumulation in seafood that can cause a variety of shellfish poisoning syndromes in humans (Wang, 2008). When chemical substances or toxic compounds are ingested by herbivorous feeders, a sort of "cascade effect" on the trophic pathway happens (Granéli et al., 1993), with repercussions across the marine food web (Delgado and Alcaraz, 1999; Teegarden and Cembella, 1996; White, 1981) and in all the trophic levels.

Reduction in HABs could be possible when zooplankton grazers increase, as happens with the rise of temperature (O'Connor et al., 2009; Sommer and Lengfellner, 2008). Grazers could lead to a reduction in the magnitude, frequency, and duration of HABs, but the other scenario may see an increase of non-HAB phytoplankton species grazers that replace HAB-grazers, allowing phytoplankton species to develop (Brandenburg et al., 2019).

Not only toxic algal blooms may affect the environment, but also non-toxic blooms. The uncontrolled growth of phytoplankton may affect marine populations in indirect effects such as anoxia, due to a major reduction in oxygen concentration (Zingone and Enevoldsen, 2000), mucilage formation, or the existence of anti-grazing chemical or morphological defenses (Delgado and Alcaraz, 1999).

A case of mass mortality of cultured mussels, *Mytilus galloprovincialis*, was detected during the winter of 1994–1995 as a consequence of the proliferation of dinoflagellates belonging to the genus *Gyrodinium* in Alfacs Bay (Ebre River delta, NW Mediterranean). The dinoflagellate was identified as *Gyrodinium corsicum*, a toxic organism responsible for the mass mortality of the copepod predator *Acartia grani*. Noxious effects of *G. corsicum* on *A. grani* occurred through direct contact (attachment of the dinoflagellate to the urosome and antennae of the copepod), disrupting the mechanical and chemical sensory systems of *A. grani*, or through indirect contact (dinoflagellate cells with their exudate filaments attached to the urosome), provoking disturbance of the antennal sensory system or clogging of the genital pore and anus. In conclusion, the mortality of *A. grani* is attributed to the dinoflagellate, although it depends on its concentration (Delgado and Alcaraz, 1999).

Phytoplankton blooms can even cause directly or indirectly the death of billions of finfish and shellfish in events dating back thousands of years (Burkholder, 1998). This proves the opinion of the scientific community during the past decades: the uncontrolled growth of marine microalgae is harmful to many marine organisms (Hallegraeff, 1993; Shumway et al., 1995; Smayda, 1990).

HABs are dangerous to humans because they are responsible for many diseases. Paralytic shellfish poisoning (PSP; from saxitoxin and its derivatives) is the most widespread form of shellfish contamination caused by harmful dinoflagellates (Hallegraeff, 1993), followed by diarrhetic shellfish poisoning (DSP; from okadaic acid and other toxins), neurotoxic shellfish poisoning (NSP; from brevetoxins), and amnesic shellfish poisoning (ASP; from domoic acid), which is caused by benthic diatoms (Bacillariophyceae) (Fritz et al., 1992; Lundholm et al., 1994; Wright et al., 1989).

Another common disease is ciguatera finfish poisoning (CFP), well documented in the United States, where intoxication has been associated with the presence of dinoflagellates producing toxins similar to those of certain marine blue-green algae (Cyanophyceae or Cyanobacteria, such as *Trichodesmium erythraeum*) (Endean et al., 1993). Some human health problems are linked to the bioaccumulation of toxins in organisms; for that reason, even consumption of small amounts of carnivorous gastropods, crustaceans (e.g., American lobster, *Homarus americanus*), and finfish (e.g., species of snapper, barracuda, grouper, and mackerel, whose consumption causes CFP) increases the risk of exposure to endotoxins (Lewis and Ruff, 1993; Shumway et al., 1995; White, 1981). The Chrysophyta *Chrysochromulina polylepis*, *Phaeocystis* sp., *Prymnesium patelliferum*, *Prymnesium parvum*, *Chattonella antiqua*, *Chattonella marina*, *Heterosigma akashiwo*, and the blue-green alga *Nodularia spumigena* are potentially lethal, and their proliferation in wild fish culture may provoke suffocation of organisms caused by the release of exotoxins and/or endotoxins (Maestrini and Graneli, 1991; Sivonen et al., 1989) or by the production of copious mucilage (Chang, 1983; Estep and MacIntyre, 1989; Savage, 1932).

Concerning the depletion of oxygen caused by algal bloom, Garcia and Johnstone (2006) investigated the impact of filamentous Cyanobacteria *Lyngbya majuscula* blooms on the meiobenthos. The decreasing level of oxygen in sediments caused a decrease in abundance and in their sediment depth distribution (Zeppilli et al., 2015). Nevertheless, the bloom did strongly affect copepod and nematode abundance and distribution, but the abundance and distribution of polychaetes were not affected as much.

During the last two decades, the most widespread HABs have been detected in the Mediterranean, particularly along the coasts of Algeria, France, Italy, and Tunisia. *Ostreopsis* sp. blooms have caused several human health problems, such as respiratory disorders (rhinorrhea), eye and nose irritation, and fever (Barroso García et al., 2008; Illoul et al., 2012; Tichadou et al., 2010; Tubaro et al., 2011; Vila et al., 2008). Even exposure to contaminated aerosols containing biotoxins or allergenic substances produced by fragmented cells of *Ostreopsis* sp. (Casabianca et al., 2013; Ciminiello et al., 2014) can be harmful to human health (Berdalet et al., 2016). In Antarctica, the composition of phytoplankton communities determines changes in carbon cycling within oceanic systems (Guglielmo et al., 2011).

Considerable increases in phytoplankton biomass and large size structure (micro-fraction accounting for 75% of the total biomass on average, independent of the functional groups) suggest that the Ross Sea in summer could now be extremely productive and might have a powerful impact on the trophic structure of the entire ecosystem (e.g., alterations in herbivore sizes, shifts in pelagic-benthic coupling, occupancy of big mammals, change in penguin diet, and others), as pointed out in previous studies (Xavier et al., 2016). The Antarctic trophic chain has seen a shift during the 21st century; while before 2000, the offshore waters were characterized by low concentrations of phytoplankton biomass, after 2014, the dominance of phytoplankton along the ice edge increased dramatically, showing a reduction of grazing activity as a proxy of scarce trophic efficiency (Mangoni et al., 2017). This shift in the abundance of phytoplankton is due to the rise in temperature in the ice environment, leading to a different nutritional quality of the phytoplankton, thus a different response in trophic web efficiency. The inefficient grazing on phytoplankton would be linked to the phytoplankton functional groups, composed of non-palatable or low nutritional value matter, enhanced in *Phyllophora antarctica* (phylum Rhodophyta) bloom by the presence of diatom accumulation (Mangoni et al., 2017).

Murphy et al. (2020) examined the combined effects of warming and acidification of seawater related to the presence of phytoplankton. There is several evidence about how climate change, mainly high temperature and decreased pH, impacts planktonic organisms through direct effects, changing physiology and behavior that can alter growth, body size, reproduction, and survival (Cripps et al., 2014; Garzke et al., 2015; Kroeker et al., 2010; Sommer et al., 2017; Weydmann et al., 2012). Still, indirect effects due to the impact given by trophic interaction chains of plankton-based food webs such as phytoplankton (Gaedke et al., 2010) affect the intermediate trophic groups more strongly, altering top-down interactions.

Warming and acidification alter phytoplankton taxonomic composition, changing also the strength of zooplankton–phytoplankton interactions, enhancing a shift from less edible (cyanobacteria) to more edible (diatoms) phytoplankton taxa in the communities studied by Murphy et al. (2020). Nevertheless, increasing temperature also reported a decrease in the size of marine phytoplankton (Sommer et al., 2017), bringing a weak trophic connection (Boyce et al., 2015). As a result, small phytoplankton cell size affects the efficiency of grazers feeding. The experiments show that warming has direct impacts on the top of plankton food webs and that top-down effects are stronger in shaping the plankton community through cascading interactions. Alterations in the food chain from indirect effects of warming, acidification, and predator decline may be equally consequences of direct effects under climate change.

To understand clearly the effect of climate change, we need to understand the effect on trophic interactions through multitrophic studies of plankton communities and study the direct and indirect effects of climate change stressors on them, to fully realize the consequences that climate change will have on aquatic ecosystems.

Through the detection of climate change, it is possible to know how populations will behave over time and to predict the structure of the community by change. To achieve this goal, it could be useful to have omic tools to provide us with information about the environment. Omic tools work better in terms of chemicals and contaminants than climate change. Nevertheless, omic techniques, such as genomics, transcriptomics, proteomics, and metabolomics, have been developed in different aspects, such as factors related to climate change, like the study of HABs.

In the last two decades, omic tools have become more popular, and a huge number of prokaryotic and eukaryotic genome sequences are already available in databases. Even though many algal genomes are already recognized, datasets are available only for limited species, such as the freshwater green alga, *Chlamydomonas reinhardtii*, which is the best-studied species, with the whole genome already sequenced (Mishra et al., 2019). Moreover, technology has been developed that has made omic datasets available, revealing mechanisms related to HAB formation and toxic production (Hennon and Dyhrman, 2020).

Omic tools may improve predictions of how global warming will impact the structure and development of HABs. Since all analytical techniques are inappropriate for all species that shape HABs, omic approaches (genomics, transcriptomics, proteomics, and metabolomics) may include a new series of tools aimed at understanding HAB dynamics (Hennon and Dyhrman, 2020). These tools are used to measure target molecules to quantify the abundance of specific genes or toxins that may be released by HAB species and support responses to environmental change (Hennon and Dyhrman, 2020). HAB responses to climate change are not easy to predict because of interference from other factors such as nutrient limitation, which reduces growth, different strains of species, adaptation, and acclimation. However, from multiple studies on HABs and climate change, it is clearly proven that many HAB species have increased growth rates or toxicity production in response to climate change perturbations (Hennon and Dyhrman, 2020).

Omic-based techniques are involved in the discovery of predictable environmental stressors through the presence of genetically determined mechanisms of HABs. Predictive responses of species to climate change are a crucial point, especially for those species that do not completely behave similarly in laboratory experiments.

Hennon and Dyhrman (2020) provide a selection of omic techniques related to HAB genera, since in the last decade, the number of omic datasets has rapidly increased. With the reduction of the cost per base in sequencing technology, there has been an increase in the number of genomes and transcriptomes. The four omic tools taken into consideration are genomics, transcriptomics, proteomics, and metabolomics.

7.6.1 GENOMICS

Genomes associated with toxin production (Pearson et al., 2016) have already been identified in Cyanobacteria *Microcystis aeruginosa* (Harke et al., 2016), *Cylindrospermopsis* (Stucken et al., 2010), and *Nodularia* (Popin et al., 2016; Voß et al., 2013), because their relatively small genome sizes facilitate genome sequencing.

For eukaryotic HAB genomes, there is already a library containing sequences of *Alexandrium*, *Prorocentrum*, and *Heterocapsa* (Jaeckisch et al., 2011; McEwan et al., 2008; Ponmani et al., 2016), as well as the haptophyte *Chrysochromulina* (John et al., 2010), used as references for understanding eukaryotic HAB toxins and genome structures. Genome sequences combined with gene expression data may provide useful insights for studying toxicity in the environment, as demonstrated by the genes involved in the synthesis of the domoic acid toxin in the *Pseudonitzschia* genome (Brunson et al., 2018).

The first species with an entire genome sequence was *Aureococcus anophagefferens* (class Pelagophyceae) (Gobler et al., 2011), which enabled the elucidation of mechanisms of bloom formation in estuaries where it occurs (Gobler and Sunda, 2012).

7.6.2 TRANSCRIPTOMICS

Since 2014, transcriptomics has been one of the most widely utilized methods for studying HABs (Hennon and Dyhrman, 2020), and the Marine Microbial Eukaryote Transcriptome Sequencing Project (MMETSP) (Keeling et al., 2014) is a database that contains hundreds of RNA sequences for eukaryotes, including over 150 belonging to HAB genera. MMETSP is a valuable source for examining gene diversity among HAB species (Di Dato et al., 2015; Liu et al., 2016; Murray et al., 2012). Through this database, it is possible to identify links between gene expression and the effects of nutrient reduction (Frischkorn et al., 2014; Haley et al., 2017; Harke et al., 2017) or increased CO_2 (Hennon et al., 2019).

Microarrays are typical tools used for transcriptomic analysis and have been employed to measure changes in response to environmental stresses in HABs of dinoflagellates *Karenia* (Morey et al., 2011; Van Dolah et al., 2007) and *Alexandrium* (Wohlrab et al., 2010; Yang et al., 2011), as well as the cyanoHAB *Microcystis* (Makower et al., 2015; Sandrini et al., 2015; Straub et al., 2011).

Other examples of the use of transcriptome approaches include the identification of genes related to saxitoxin production in the genus *Alexandrium* (Stüken et al., 2011) and the gene candidates for the production of ciguatoxins in *Gambierdiscus polynesiensis* (phylum Miozoa) (Pawlowiez et al., 2014), providing assurance that these tools will yield important information regarding the biology and response of HAB species to environmental changes (Hennon and Dyhrman, 2020).

7.6.3 PROTEOMICS

Another key tool is proteomics, which is fundamental for measuring protein abundance and detecting enzymatic activity (Chan et al., 2004; Zhang et al., 2015). Nowadays, the wide availability of peptide sequences in databases allows scientists to understand changes in peptide abundance and use this information as a proxy to detect proteins in the environment. Current proteomics studies have identified the proteomes of HAB genera such as *Microcystis* (Cyanophyceae) (Chen et al., 2017; Yue et al., 2015), *Prorocentrum*, *Alexandrium* (Dinophyceae) (Wang et al., 2012, 2013), *Aureoumbra* (Sun et al., 2012), and *Aureococcus* (Pelagophyceae) (Wurch et al., 2011), which show changes in physiology in response to environmental stressors.

7.6.4 METABOLOMICS

The least used omic tool is metabolomics. These techniques are growing in use, especially for the characterization of toxic metabolites (Briand et al., 2016), but they may also be used to evaluate cellular responses to shifting environments, assessing osmotic stress and nutrient reduction (Hennon and Dyhrman, 2020). Metabolomics may reveal environmental signals due to interactions between bacteria and HAB species (Amin et al., 2015) or allelopathic correlations with competing algae (Poulin et al., 2018; Song et al., 2017). The two omic tools, metabolomics and proteomics, were used to determine the impact of *Karenia brevis* allelopathy on the diatom species *Thalassiosira pseudonana*, which showed that the growth rate had a drastic reduction in co-culture with *K. brevis*. Moreover, proteomes indicated a decrease in photosynthesis and an increase in oxidative stress (Poulson-Ellestad et al., 2014).

HABs are clearly affected by global warming, as many studies have suggested (Dale et al., 2006; Hennon and Dyhrman, 2020; Papiol, 2018; Tester et al., 2020; Wells et al., 2015). Omic techniques combined could provide delineated responses to environmental changes brought about by phytoplankton abundance. Although there is still a lack of knowledge about genome, transcriptome, proteome, and metabolome datasets for HAB species, researchers are still working on it to develop strong tools that could be essential in improving the prediction of HAB events in the face of climate change (Hennon and Dyhrman, 2020).

Since microalgae have a great number of associated advantages, it is important to preserve them. Among the potential uses of microalgae, there is the production of biofuel and other co-products, thanks to their ability to produce and store energy and their high amounts of proteins, carbohydrates, and oil. This is an advantage even for the environment: having renewable fuel instead of fossil fuel and coal (Mishra et al., 2019). Moreover, microalgae are good at sequestering CO_2, and they act as bioremediators, utilizing nutrients like phosphate and nitrogen and remediating pollutants from water bodies like sewage, wastewater, and industrial effluents (Kumar et al., 2011; Park et al., 2011). In addition, micronutrients present in microalgae could be used as food supplements as well as animal feeds (Camacho et al., 2019; Smerilli et al., 2017).

7.7 CONCLUSION

Global warming is a threat that is likely to intensify in the coming years, and its effects are dangerous for the marine environment and ecosystems. Under the effects of climate change, represented by a rise in temperature, an increase in CO_2, OA, and changes in pH and salinity, a wide variety of marine organisms have been strongly affected in their functioning, behavior, and distribution.

The present work focuses on the consequences of climate change on microalgae. Studies have shown that global warming affects microalgae, particularly their abundance, and enhances HABs. Microalgal blooms are a natural seasonal phenomenon in marine ecosystems around the world. Nonetheless, blooms are greatly influenced by climate change and related factors. On the other hand, the adaptability and plasticity of microalgae allow some strains to adapt and acclimate under nonoptimal conditions, such as increased CO_2, and changes in temperature, salinity, and pH. Collins and Bell (2004) discovered that under stress conditions caused by increased CO_2, microalgae can transfer genetic information acquired through acclimation to their offspring; this means that the offspring of acclimated species have a greater chance of surviving environmental changes associated with global warming.

Climate change is closely related to the presence of HABs because its effects have repercussions on their abundance and distribution, altering their seasonality and causing severe impacts through the release of toxins or the occurrence of hypoxia, which can lead to the death of marine organisms such as zooplankton and fish. Moreover, environmental changes that affect microalgae lead to a "cascade effect" in all the functions of organisms involved in trophic webs that include microalgae. For this reason, it is important to protect microalgae, as they play a pivotal role at the base of marine ecosystems and food webs. Additionally, microalgae are important to preserve because they represent a significant resource in various fields, including pharmaceutical, nutraceutical, and bioremediation applications. Some species of algae have already been identified as useful for the bioremediation of toxic substances released into the environment; therefore, it is necessary to preserve them and prevent changes in their structure and physiology due to climate change. During the last decade, researchers have developed omics tools using microalgae to predict the impacts caused by climate change. Nevertheless, there is limited information on this topic, but scientists continue to collect additional data to predict HAB trends and to understand how to prevent changes in distribution and loss of biodiversity.

ACKNOWLEDGMENTS

This work was funded by European funds through the European Regional Development Fund (FEDER), under the Centro 2030 Programme, project

"MARCentro+ - Inovação e Sustentabilidade na Gestão dos Recursos Marinhos da Região Centro" (CENTRO2030-FEDER-02614400). This work was supported by FCT - Fundação para a Ciência e Tecnologia, I.P., in the framework of the Project UIDB/04004/2025 - Centre for Functional Ecology - Science for the People & the Planet, financed by FCT/MCTES through national funds (PIDDAC), Associate Laboratory TERRA (LA/P/0092/2020; https://doi.org/10.54499/LA/P/0092/2020) and the project UIDB/04292/2020-MARE-Marine and Environmental Sciences Centre. Silvia Lomartire acknowledges the project PORBIOTA—E-Infrastructure Portuguese Information and Research in Biodiversity (POCI-01-0145-FEDER-022127), which financed this research and was supported by the Competitiveness and Internationalization Operational Program and the Regional Operational Program of Lisbon, through FEDER, and by the Portuguese Foundation for Science and Technology (FCT), through national funds (OE). Sara García-Poza acknowledges the project MENU—Marine Macroalgae: Alternative Recipes for a Daily Nutritional Diet (FA_05_2017_011), which partially financed this research and was funded by the Blue Fund under Public Notice No. 5—Blue Biotechnology. Diana Pacheco acknowledges PTDC/BIA-CBI/31144/2017—POCI-01-0145-FEDER-031144—MARINE INVADERS, cofinanced by the European Regional Development Fund (ERDF) through POCI (Operational Program Competitiveness and Internationalization) and by the Foundation for Science and Technology (FCT, IP).

KEYWORDS

- **bioremediation**
- **carbon dioxide**
- **environmental assessment**
- **inorganic carbon transport**
- **marine organisms**
- **omic tools**
- **phytoplankton biomass**
- **thermotolerant microalgae**
- **trophic chain**

REFERENCES

Abdel-Ghani, N. T., & El-Chaghaby, G. A. (2014). Biosorption for metal ions removal from aqueous solutions: A review of recent studies. *International Journal of Latest Research in Science and Technology, 3*(1), 24–42.

Abinandan, S., Suresh, R. S., Kadiyala, V., & Mallavarapu, M. (2018). Nutrient removal and biomass production: Advances in microalgal biotechnology for wastewater treatment. *Critical Reviews in Biotechnology, 38*(8), 1244–1260. https://doi.org/10.1080/07388551.2018.1472066

Abu-Ghosh, S., Pal-Nath, D., Markovitch, D., Solovchenko, A., Didi-Cohen, S., Portugal, I., et al. (2015). A novel source of dihomo-γ-linolenic acid: Possibilities and limitations of DGLA production in high-density cultures of the Δ5 desaturase-mutant microalga *Lobosphaera incisa*. *European Journal of Lipid Science and Technology, 117*(6), 760–766. https://doi.org/10.1002/ejlt.201400430 (accessed on 20th February 2026).

Abu-Ghosh, S., Zvy, D., & David, I. (2020). Acclimation of thermotolerant algae to light and temperature interaction. *Journal of Phycology*, 1–3. https://doi.org/10.1111/jpy.12964 (accessed on 20th February 2026).

Acevedo-Trejos, E., Brandt, G., Steinacher, M., & Merico, A. (2014). A glimpse into the future composition of marine phytoplankton communities. *Frontiers in Marine Science, 1*(7), 1–12. https://doi.org/10.3389/fmars.2014.00015 (accessed on 20th February 2026).

Adar, O., Kaplan-Levy, R. N., & Banet, G. (2016). High temperature Chlorellaceae (Chlorophyta) strains from the Syrian African rift valley: The effect of salinity and temperature on growth, morphology and sporulation mode. *European Journal of Phycology, 51*(4), 387–400. https://doi.org/10.1080/09670262.2016.1193772 (accessed on 20th February 2026).

Alda-Vidal, C., Browne, A. L., & Hoolohan, C. (2020). "Unflushables": Establishing a global agenda for action on everyday practices associated with sewer blockages, water quality, and plastic pollution. *Wiley Interdisciplinary Reviews: Water, 7*(4), e1452.

Amin, S. A., Hmelo, L. R., van Tol, H. M., Durham, B. P., Carlson, L. T., Heal, K. R., et al. (2015). Interaction and signaling between a cosmopolitan phytoplankton and associated bacteria. *Nature, 522*, 98–101. https://doi.org/10.1038/nature14488 (accessed on 20th February 2026).

Anderson, D. M., Cembella, A. D., & Hallegraeff, G. M. (2012). Progress in understanding harmful algal blooms: Paradigm shifts and new technologies for research, monitoring, and management. *Annual Review of Marine Science, 4*(1), 143–176. https://doi.org/10.1146/annurev-marine-120308-081121 (accessed on 20th February 2026).

Anderson, D. M., Glibert, P. M., & Burkholder, J. M. (2002). Harmful algal blooms and eutrophication: Nutrient sources, composition, and consequences. *Estuaries, 25*(4), 704–726.

Aoki, M., Suematsu, H., Kumata, H., & Fujiwara, K. (2008). Physiological and photosynthetic toxicity of thallium in *Synechocystis sp.* PCC6803. In A. Allen et al. (Eds.), *Photosynthesis: Energy from the Sun* (pp. 1399–1402). Springer, Netherlands.

Arce, O., Herbas, R., Rivero, F., & González, A. (2006). *Indicadores Biológicos de Calidad del Agua.* Universidad Mayor de San Simón.

Archer, D., Winguth, A., Lea, D., & Mahowald, N. (2000). What caused the glacial/interglacial atmospheric pCO_2 cycles? *Reviews of Geophysics, 38*(2), 159–189.

Ardal, E. (2014). *Phycoremediation of Pesticides Using Microalgae.* Swedish University of Agricultural Sciences.

Baba, M., & Shiraiwa, Y. (2012). High-CO_2 response mechanisms in microalgae. In *Advances in Photosynthesis: Fundamental Aspects*. https://doi.org/10.5772/27737 (accessed on 20th February 2026).

Badger, M. R., Whitney, S. M., Ludwig, M., Yellowlees, D. C., Leggat, W., & Price, G. D. (1998). The diversity and coevolution of RuBisCO, plastids, pyrenoids, and chloroplast-based CO_2-concentrating mechanisms in algae. *Canadian Journal of Botany, 76*(6), 1052–1071. https://doi.org/10.1139/b98-074 (accessed on 20th February 2026).

Baker, A. C., Starger, C. J., McClanahan, T. R., & Glynn, P. W. (2004). Coral's adaptive response to climate change. *Nature, 430*(7001), 741. https://doi.org/10.1038/430741a (accessed on 20th February 2026).

Baltar, F., Bayer, B., Bednarski, N., Deppeler, S., Escribano, R., Gonzalez, C. E., et al. (2019). Towards integrating evolution, metabolism, and climate change studies of marine ecosystems. *Trends in Ecology & Evolution, 34*(11), 1022–1033. https://doi.org/10.1016/j.tree.2019.07.003 (accessed on 20th February 2026).

Balzano, S., Sardo, A., Blasio, M., Bou Chahine, T., Dell'Anno, F., Sansone, C., et al. (2020). Microalgal metallothioneins and phytochelatins and their potential use in bioremediation. *Frontiers in Microbiology, 11*, 517. https://doi.org/10.3389/fmicb.2020.00517 (accessed on 20th February 2026).

Barange, M., Merino, G., Blanchard, J. L., Scholtens, J., Harle, J., Allison, E. H., et al. (2014). Impacts of climate change on marine ecosystem production in societies dependent on fisheries. *Nature Climate Change, 4*(3), 211–216. https://doi.org/10.1038/nclimate2119 (accessed on 20th February 2026).

Barreiro, A., Carotenuto, Y., Lamari, N., Esposito, F., D'Ippolito, G., Fontana, A., et al. (2011). Diatom induction of reproductive failure in copepods: The effect of PUAs versus non-volatile oxylipins. *Journal of Experimental Marine Biology and Ecology, 401*(1, 2), 13–19. https://doi.org/10.1016/j.jembe.2011.03.007 (accessed on 20th February 2026).

Barroso, G. P., Rueda de la Puerta, P., Carreño, T. P., Martínez, P. M., & Enríquez, J. G. (2008). Brote con síntomas respiratorios en la provincia de Almería por una posible exposición a microalgas tóxicas. *Gaceta Sanitaria, 22*(6), 578–584. https://doi.org/10.1016/S0213-9111(08)75357-3 (accessed on 20th February 2026).

Bautista-Chamizo, E., Sendra, M., Cid, Á., Seoane, M., Romano de Orte, M., & Riba, I. (2018). Will temperature and salinity changes exacerbate the effects of seawater acidification on the marine microalga *Phaeodactylum tricornutum*? *Science of the Total Environment, 634*, 87–94. https://doi.org/10.1016/j.scitotenv.2018.03.314 (accessed on 20th February 2026).

Beardall, J., & Raven, J. A. (2004). The potential effects of global climate change on microalgal photosynthesis, growth and ecology. *Phycological, 43*(1), 26–40. https://doi.org/10.2216/i0031-8884-43-1-26.1 (accessed on 20th February 2026).

Beardall, J., & Stojkovic, S. (2006). Microalgae under global environmental change: Implications for growth and productivity, populations and trophic flow. *Science Asia, 32*(S1), 1–10. https://doi.org/10.2306/scienceasia1513-1874.2006.32(s1).001 (accessed on 20th February 2026).

Beardall, J., Heraud, P., Roberts, S., Shelly, K., & Stojkovic, S. (2002). Effects of UV-B radiation on inorganic carbon acquisition by the marine microalga *Dunaliella tertiolecta*

(Chlorophyceae). *Phycological, 41*(3), 268–272. https://doi.org/10.2216/i0031-8884-41-3-268.1 (accessed on 20th February 2026).

Behrenfeld, M. J., Lean, D. R. S., & Lee, H. (1995). Ultraviolet-B radiation effects on inorganic nitrogen uptake by natural assemblages of oceanic plankton. *Journal of Phycology, 31*(1), 25–36. https://doi.org/10.1111/j.0022-3646.1995.00025..x (accessed on 20th February 2026).

Berdalet, E., Fleming, L. E., Gowen, R., Davidson, K., Hess, P., Backer, L. C., et al. (2016). Marine harmful algal blooms, human health and wellbeing: Challenges and opportunities in the 21st century. *Journal of the Marine Biological Association of the United Kingdom, 96*(1), 61–91. https://doi.org/10.1017/S0025315415001733 (accessed on 20th February 2026).

Berge, T., Daugbjerg, N., Andersen, B. B., & Hansen, P. J. (2010). Effect of lowered pH on marine phytoplankton growth rates. *Marine Ecology Progress Series, 416*, 79–91. https://doi.org/10.3354/meps08780 (accessed on 20th February 2026).

Berry, J. A., & Raison, J. K. (1981). Responses of macrophytes to temperature. In O. L. Lange et al. (Eds.), *Physiological Plant Ecology* (pp. 277–338). https://doi.org/10.1007/978-3-642-68090-8_11 (accessed on 20th February 2026).

Berry, J., & Bjorkman, O. (1980). Photosynthetic response and adaptation to temperature in higher plants. *Annual Review of Plant Physiology, 31*(1), 491–543. https://doi.org/10.1146/annurev.pp.31.060180.002423 (accessed on 20th February 2026).

Bilal, M., Rasheed, T., Sosa-Hernández, J., Raza, A., Nabeel, F., & Iqbal, H. (2018). Biosorption: An interplay between marine algae and potentially toxic elements—A review. *Marine Drugs, 16*(2), 65. https://doi.org/10.3390/md16020065 (accessed on 20th February 2026).

Blanchard, J. L., Jennings, S., Holmes, R., Harle, J., Merino, G., Allen, J. I., et al. (2012). Potential consequences of climate change for primary production and fish production in large marine ecosystems. *Philosophical Transactions of the Royal Society B: Biological Sciences, 367*, 2979–2989. https://doi.org/10.1098/rstb.2012.0231 (accessed on 20th February 2026).

Blanchard, J. L., Watson, R. A., Fulton, E. A., Cottrell, R. S., Nash, K. L., Bryndum-Buchholz, A., et al. (2017). Linked sustainability challenges and trade-offs among fisheries, aquaculture and agriculture. *Nature Ecology & Evolution, 1*(9), 1240–1249. https://doi.org/10.1038/s41559-017-0258-8 (accessed on 20th February 2026).

Bopp, L., Resplandy, L., Orr, J. C., Doney, S. C., Dunne, J. P., Gehlen, M., et al. (2013). Multiple stressors of ocean ecosystems in the 21st century: Projections with CMIP5 models. *Bio Geosciences, 10*(10), 6225–6245. https://doi.org/10.5194/bg-10-6225-2013 (accessed on 20th February 2026).

Borges, A. V., & Gypens, N. (2010). Carbonate chemistry in the coastal zone responds more strongly to eutrophication than to ocean acidification. *Limnology and Oceanography, 55*(1), 346–353. https://doi.org/10.4319/lo.2010.55.1.0346 (accessed on 20th February 2026).

Boudh, S., & Singh, J. S. (2018). Pesticide contamination: Environmental problems and remediation strategies. In *Emerging and Eco-Friendly Approaches for Waste Management* (pp. 245–269). Springer, Singapore.

Boyce, D. G., Frank, K. T., & Leggett, W. C. (2015). From mice to elephants: Overturning the "one size fits all" paradigm in marine plankton food chains. *Ecology Letters, 18*(6), 504–515. https://doi.org/10.1111/ele.12434 (accessed on 20th February 2026).

Boyle, N. R., & Morgan, J. A. (2009). Flux balance analysis of primary metabolism in *Chlamydomonas reinhardtii*. *BMC Systems Biology, 3*, 4. https://doi.org/10.1186/1752-0509-3-4 (accessed on 20th February 2026).

Brandenburg, K. M., Velthuis, M., & Van de Waal, D. B. (2019). Meta-analysis reveals enhanced growth of marine harmful algae from temperate regions with warming and elevated CO_2 levels. *Global Change Biology, 25*(8), 2607–2618. https://doi.org/10.1111/gcb.14678 (accessed on 20th February 2026).

Briand, E., Bormans, M., Gugger, M., Dorrestein, P. C., & Gerwick, W. H. (2016). Changes in secondary metabolic profiles of *Microcystis aeruginosa* strains in response to intraspecific interactions. *Environmental Microbiology, 18*(2), 384–400. https://doi.org/10.1111/1462-2920.12904 (accessed on 20th February 2026).

Brodie, J., Andersen, R. A., Kawachi, M., & Millar, A. J. K. (2009). Endangered algal species and how to protect them. *Phycological, 48*(5), 423–438. https://doi.org/10.2216/09-21.1 (accessed on 20th February 2026).

Brueggeman, A. J., Gangadharaiah, D. S., Cserhati, M. F., Casero, D., Weeks, D. P., & Ladunga, I. (2012). Activation of the carbon concentrating mechanism by CO_2 deprivation coincides with massive transcriptional restructuring in *Chlamydomonas reinhardtii*. *Plant Cell, 24*(5), 1860–1875. https://doi.org/10.1105/tpc.111.093435 (accessed on 20th February 2026).

Brunson, J. K., McKinnie, S. M. K., Chekan, J. R., McCrow, J. P., Miles, Z. D., Bertrand, E. M., et al. (2018). Biosynthesis of the neurotoxin domoic acid in a bloom-forming diatom. *Science, 361*(6409), 1356–1358. https://doi.org/10.1126/science.aau0382 (accessed on 20th February 2026).

Buma, A. G. J., Wright, S. W., Van den Enden, R., Van de Poll, W. H., & Davidson, A. T. (2006). PAR acclimation and UVBR-induced DNA damage in Antarctic marine microalgae. *Marine Ecology Progress Series, 315*, 33–42. https://doi.org/10.3354/meps315033 (accessed on 20th February 2026).

Bunt, J. S. (1964). Primary productivity under sea ice in Antarctic waters: I. Concentrations and photosynthetic activities of microalgae in the waters of McMurdo Sound, Antarctica. *Antarctic Research Series, 1*, 13–26. https://doi.org/10.1029/AR001p0013 (accessed on 20th February 2026).

Burkholder, J. A. M. (1998). Implications of harmful microalgae and heterotrophic dinoflagellates in management of sustainable marine fisheries. *Ecological Applications, 8(1), S37–S62*. https://doi.org/10.2307/2641362 (accessed on 20th February 2026).

Caldeira, K., & Wickett, M. E. (2003). Anthropogenic carbon and ocean pH. *Nature, 425*(6956), 365. https://doi.org/10.1038/425365a (accessed on 20th February 2026).

Camacho, F., Macedo, A., & Malcata, F. (2019). Potential industrial applications and commercialization of microalgae in the functional food and feed industries: A short review. *Marine Drugs, 17*(6), 312. https://doi.org/10.3390/md17060312 (accessed on 20th February 2026).

Casabianca, S., Casabianca, A., Riobó, P., Franco, J. M., Vila, M., & Penna, A. (2013). Quantification of the toxic dinoflagellate *Ostreopsis* spp. by qPCR assay in marine aerosol. *Environmental Science & Technology, 47*(8), 3788–3795. https://doi.org/10.1021/es305018s (accessed on 20th February 2026).

Chan, L. L., Hodgkiss, I. J., Lu, S., & Lo, S. C. L. (2004). Use of two-dimensional gel electrophoresis proteome reference maps of dinoflagellates for species recognition of causative agents of harmful algal blooms. *Proteomics, 4*(1), 180–192. https://doi.org/10.1002/pmic.200300548 (accessed on 20th February 2026).

Chang, F. H. (1983). The mucilage-producing *Phaeocystis pouchetii* (Prymnesiophyceae), cultured from the 1981 "Tasman Bay Slime". *New Zealand Journal of Marine and Freshwater Research, 17*(2), 165–168. https://doi.org/10.1080/00288330.1983.9515994 (accessed on 20th February 2026).

Chang, R. L., Ghamsari, L., Manichaikul, A., Hom, E. F. Y., Balaji, S., Fu, W., et al. (2011). Metabolic network reconstruction of *Chlamydomonas* offers insight into light-driven algal metabolism. *Molecular Systems Biology, 7*, 518. https://doi.org/10.1038/msb.2011.52 (accessed on 20th February 2026).

Chen, H., Zhang, Z., Tian, F., Zhang, L., Li, Y., Cai, W., et al. (2018). The effect of pH on the acute toxicity of phenanthrene in the marine microalga *Chlorella salina. Scientific Reports, 8*(1), 1–8. https://doi.org/10.1038/s41598-018-35686-9 (accessed on 20th February 2026).

Chen, S., Liu, Y., Zhang, J., & Gao, B. (2017). iTRAQ-based quantitative proteomic analysis of *Microcystis aeruginosa* exposed to plicamycin at different nutrient levels. *Aquatic Toxicology, 185*, 193–200. https://doi.org/10.1016/j.aquatox.2017.02.015 (accessed on 20th February 2026).

Cheung, W. W. L., Lam, V. W. Y., Sarmiento, J. L., Kearney, K., Watson, R., Zeller, D., et al. (2010). Large-scale redistribution of maximum fisheries catch potential in the global ocean under climate change. *Global Change Biology, 16*(1), 24–35. https://doi.org/10.1111/j.1365-2486.2009.01995.x (accessed on 20th February 2026).

Chowdhury, S., Mazumder, M. A. J., Al-Attas, O., & Husain, T. (2016). Heavy metals in drinking water: Occurrences, implications, and future needs in developing countries. *Science of the Total Environment, 569, 570*, 476–488.

Chust, G., Allen, J. I., Bopp, L., Schrum, C., Holt, J., Tsiaras, K., et al. (2014). Biomass changes and trophic amplification of plankton in a warmer ocean. *Global Change Biology, 20*(7), 2124–2139. https://doi.org/10.1111/gcb.12562 (accessed on 20th February 2026).

Ciminiello, P., Dell'Aversano, C., Dello Iacovo, E., Fattorusso, E., Forino, M., Tartaglione, L., et al. (2014). First finding of *Ostreopsis ovata* toxins in marine aerosols. *Environmental Science & Technology, 48*(7), 3532–3540.

Collins, M. (2007). Ensembles and probabilities: A new era in the prediction of climate change. *Philosophical Transactions of the Royal Society A: Mathematical, Physical and Engineering Sciences, 365*(1857), 1957–1970. https://doi.org/10.1098/rsta.2007.2068 (accessed on 20th February 2026).

Collins, S., & Bell, G. (2004). Phenotypic consequences of 1,000 generations of selection at elevated CO_2 in a green alga. *Nature, 431*(7008), 566–569. https://doi.org/10.1038/nature02945 (accessed on 20th February 2026).

Collins, S., Rost, B., & Rynearson, T. A. (2013). Evolutionary potential of marine phytoplankton under ocean acidification. *Evolutionary Applications, 7*(1), 140–155. https://doi.org/10.1111/eva.12120 (accessed on 20th February 2026).

Cripps, G., Lindeque, P., & Flynn, K. J. (2014). Have we been underestimating the effects of ocean acidification in zooplankton? *Global Change Biology, 20*(11), 3377–3385. https://doi.org/10.1111/gcb.12582 (accessed on 20th February 2026).

Cuellar-Bermudez, S. P., Romero-Ogawa, M. A., Vannela, R., Lai, Y. J. S., Rittmann, B. E., & Parra-Saldivar, R. (2015). Effects of light intensity and carbon dioxide on lipids and fatty acids produced by *Synechocystis sp.* PCC6803 during continuous flow. *Algal Research, 12*, 10–16. https://doi.org/10.1016/j.algal.2015.07.018 (accessed on 20th February 2026).

Dale, B., & Reid, P. C. (2006). Climate change and harmful algal blooms. In *Ecology of Harmful Algae* (pp. 367–378). https://doi.org/10.1007/978-3-540-32210-8_28 (accessed on 20th February 2026).

Delgado, M., & Alcaraz, M. (1999). Interactions between red tide microalgae and herbivorous zooplankton: The noxious effects of *Gyrodinium corsicum* (Dinophyceae) on *Acartia grani* (Copepoda: Calanoida). *Journal of Plankton Research, 21*(12), 2361–2371. https://doi.org/10.1093/plankt/21.12.2361 (accessed on 20th February 2026).

Deppeler, S., Petrou, K., Schulz, K. G., Westwood, K., Pearce, I., McKinlay, J., et al. (2018). Ocean acidification of a coastal Antarctic marine microbial community reveals a critical threshold for CO_2 tolerance in phytoplankton productivity. *Bio Geosciences, 15*(1), 209–231. https://doi.org/10.5194/bg-15-209-2018 (accessed on 20th February 2026).

Devonshire, A. L., & Field, L. M. (1991). Gene amplification and insecticide resistance. *Annual Review of Entomology, 36*, 1–23.

Di Dato, V., Musacchia, F., Petrosino, G., Patil, S., Montresor, M., Sanges, R., et al. (2015). Transcriptome sequencing of three *Pseudo-nitzschia* species reveals comparable gene sets and the presence of nitric oxide synthase genes in diatoms. *Scientific Reports, 5*, 12329. https://doi.org/10.1038/srep12329 (accessed on 20th February 2026).

Dishisha, T., Pereyra, L. P., Pyo, S. H., Britton, R. A., & Hatti-Kaul, R. (2014). Flux analysis of the *Lactobacillus reuteri* propanediol-utilization pathway for production of acid and 1,3-propanediol from glycerol. *Microbial Cell Factories, 13*(76), 1–10.

Döhler, G. (1985). Effect of UV-B radiation (290–320 nm) on the nitrogen metabolism of several marine diatoms. *Journal of Plant Physiology, 118*(5), 391–400. https://doi.org/10.1016/S0176-1617(85)80199-1 (accessed on 20th February 2026).

Doney, S. C., Ruckelshaus, M., Duffy, J. E., Barry, J. P., Chan, F., English, C. A., et al. (2012). Climate change impacts on marine ecosystems. *Annual Review of Marine Science, 4*(1), 11–37. https://doi.org/10.1146/annurev-marine-041911-111611 (accessed on 20th February 2026).

Dove, M. C., & Sammut, J. (2007). Impacts of estuarine acidification on survival and growth of Sydney rock oysters *Saccostrea glomerata* (Gould, 1850). *Journal of Shellfish Research, 26*(2), 519–527. https://doi.org/10.2983/0730-8000(2007)26[519:ioeaos]2.0.co;2 (accessed on 20th February 2026).

Duce, R. A., & Tindale, N. T. (1991). Atmospheric transport of iron and its deposition in the ocean. *Limnology and Oceanography, 36*(8), 1715–1726. https://doi.org/10.4319/lo.1991.36.8.1715 (accessed on 20th February 2026).

Durack, P. J., & Wijffels, S. E. (2010). Fifty-year trends in global ocean salinities and their relationship to broad-scale warming. *Journal of Climate, 23*, 4342–4362. https://doi.org/10.1175/2010JCLI3377.1 (accessed on 20th February 2026).

Dutkiewicz, S., Morris, J. J., Follows, M. J., Scott, J., Levitan, O., Dyhrman, S. T., et al. (2015). Impact of ocean acidification on the structure of future phytoplankton communities. *Nature Climate Change, 5*(11), 1002–1006. https://doi.org/10.1038/nclimate2722 (accessed on 20th February 2026).

Dutkiewicz, S., Scott, J. R., & Follows, M. J. (2013). Winners and losers: Ecological and biogeochemical changes in a warming ocean. *Global Biogeochemical Cycles, 27*(2), 463–477. https://doi.org/10.1002/gbc.20042 (accessed on 20th February 2026).

Edwards, M., Johns, D. G., Leterme, S. C., Svendsen, E., & Richardson, A. J. (2006). Regional climate change and harmful algal blooms in the Northeast Atlantic. *Limnology and Oceanography, 51*(2), 820–829. https://doi.org/10.4319/lo.2006.51.2.0820 (accessed on 20th February 2026).

Eerkes-Medrano, D., Leslie, H. A., & Quinn, B. (2019). Microplastics in drinking water: A review and assessment. *Current Opinion in Environmental Science & Health, 7*, 69–75.

Egilsdottir, H., Spicer, J. I., & Rundle, S. D. (2009). The effect of CO_2-acidified seawater and reduced salinity on aspects of the embryonic development of the amphipod *Echinogammarus marinus* (Leach). *Marine Pollution Bulletin, 58*(8), 1187–1191. https://doi.org/10.1016/j.marpolbul.2009.03.017 (accessed on 20th February 2026).

El-Ansari, O., & Colman, B. (2015). Inorganic carbon acquisition in the acid-tolerant alga *Chlorella kessleri*. *Physiologiae Plantarum, 153*(1), 175–182. https://doi.org/10.1111/ppl.12228 (accessed on 20th February 2026).

El-Sayed, S. Z., & Taguchi, S. (1981). Primary production and standing crop of phytoplankton along the ice-edge in the Weddell Sea. *Deep-Sea Research Part A: Oceanographic Research Papers, 28*(9), 1017–1032. https://doi.org/10.1016/0198-0149(81)90015-7 (accessed on 20th February 2026).

Emparan, Q., Harun, R., & Danquah, M. K. (2018). Role of phycoremediation for nutrient removal from wastewaters: A review. *Applied Ecology and Environmental Research, 17*(1), 889–915. https://doi.org/10.15666/aeer/1701_889915 (accessed on 20th February 2026).

Endean, R., Monks, S. A., Griffith, J. K., & Llewellyn, L. E. (1993). Apparent relationships between toxins elaborated by the cyanobacterium *Trichodesmium erythraeum* and those present in the flesh of the narrow-barred Spanish mackerel *Scomberomorus commersoni*. *Toxicon, 31*(9), 1155–1165. https://doi.org/10.1016/0041-0101(93)90131-2 (accessed on 20th February 2026).

Estep, K. W., & MacIntyre, F. (1989). Taxonomy, life cycle, distribution and dasmotrophy of *Chrysochromulina*: A theory accounting for scales, haptonema, muciferous bodies and toxicity. *Marine Ecology Progress Series, 57*, 11–21. https://doi.org/10.3354/meps057011 (accessed on 20th February 2026).

Estes, J. A., Heithaus, M., McCauley, D. J., Rasher, D. B., & Worm, B. (2016). Megafaunal impacts on structure and function of ocean ecosystems. *Annual Review of Environment and Resources, 41*(1), 83–116.

Fabry, V. J., Feely, R. A., Orr, J. C., et al. (2008). Impacts of ocean acidification on marine biodiversity. In *Biodiversity and Climate Change: Transforming the Biosphere* (pp. 185–195). https://doi.org/10.2307/j.ctv8jnzw1.25 (accessed on 20th February 2026).

Fauchot, J., Gosselin, M., Levasseur, M., Mostajir, B., Belzile, C., Demers, S., et al. (2000). Influence of UV-B radiation on nitrogen utilization by a natural assemblage of phytoplankton. *Journal of Phycology, 36*(3), 484–496. https://doi.org/10.1046/j.1529-8817.2000.99011.x (accessed on 20th February 2026).

Fielding, A. S., Turpin, D. H., Guy, R. D., Calvert, S. E., Crawford, D. W., & Harrison, P. J. (1998). Influence of the carbon concentrating mechanism on carbon stable isotope discrimination by the marine diatom *Thalassiosira pseudonana*. *Canadian Journal of*

Botany, 76(6), 1098–1103. https://doi.org/10.1139/b98-069 (accessed on 20th February 2026).

Filimonova, V., Nys, C., De Schamphelaere, K. A. C., Gonçalves, F., Marques, J. C., Gonçalves, A. M. M., et al. (2018). Ecotoxicological and biochemical mixture effects of an herbicide and a metal at the marine primary producer diatom *Thalassiosira weissflogii* and the primary consumer copepod *Acartia tonsa*. *Environmental Science and Pollution Research, 25*(22), 22180–22195. https://doi.org/10.1007/s11356-018-2302-x (accessed on 20th February 2026).

Flynn, K. J. (2010). Ecological modeling in a sea of variable stoichiometry: Dysfunctionality and the legacy of Redfield and Monod. *Progress in Oceanography, 84*(1, 2), 52–65. https://doi.org/10.1016/j.pocean.2009.09.006 (accessed on 20th February 2026).

Follows, M. J., Dutkiewicz, S., Grant, S., & Chisholm, S. W. (2007). Emergent biogeography of microbial communities in a model ocean. *Science, 315*(5820), 1843–1846. https://doi.org/10.1126/science.1138544 (accessed on 20th February 2026).

Fomina, M., & Gadd, G. M. (2014). Biosorption: Current perspectives on concept, definition and application. *Bioresource Technology, 160*, 3–14. https://doi.org/10.1016/j.biortech.2013.12.102 (accessed on 20th February 2026).

Food and Agriculture Organization of the United Nations. (2004). *The State of World Fisheries and Aquaculture.*

Free, C. M., Thorson, J. T., Pinsky, M. L., Oken, K. L., Wiedenmann, J., & Jensen, O. P. (2019). Impacts of historical warming on marine fisheries production. *Science, 363*(6430), 979–983. https://doi.org/10.1126/science.aax5721 (accessed on 20th February 2026).

Frei, C., Schöll, R., Fukutome, S., Schmidli, J., & Vidale, P. L. (2006). Future change of precipitation extremes in Europe: Intercomparison of scenarios from regional climate models. *Journal of Geophysical Research: Atmospheres, 111*(D6). https://doi.org/10.1029/2005JD005965 (accessed on 20th February 2026).

Frischkorn, K. R., Harke, M. J., Gobler, C. J., & Dyhrman, S. T. (2014). *De novo* assembly of *Aureococcus anophagefferens* transcriptomes reveals diverse responses to low nutrient and low light conditions present during blooms. *Frontiers in Microbiology, 5*, 375. https://doi.org/10.3389/fmicb.2014.00375 (accessed on 20th February 2026).

Fritz, L., Quilliam, M. A., Wright, J. L. C., Beale, A. M., & Work, T. M. (1992). An outbreak of domoic acid poisoning attributed to the pennate diatom *Pseudonitzschia australis*. *Journal of Phycology, 28*, 439–442.

Frölicher, T. L., Sarmiento, J. L., Paynter, D. J., Dunne, J. P., Krasting, J. P., & Winton, M. (2015). Dominance of the Southern Ocean in anthropogenic carbon and heat uptake in CMIP5 models. *Journal of Climate, 28*(2), 862–886. https://doi.org/10.1175/JCLI-D-14-00117.1 (accessed on 20th February 2026).

Fu, F., Warner, M. E., Zhang, Y., Feng, Y., & Hutchins, D. A. (2007). Effects of increased temperature and CO_2 on photosynthesis, growth, and elemental ratios in marine *Synechococcus* and *Prochlorococcus* (Cyanobacteria). *Journal of Phycology, 43*, 485–496. https://doi.org/10.1111/j.1529-8817.2007.00355.x (accessed on 20th February 2026).

Fuller, A., Dawson, T., Helmuth, B., Hetem, R. S., Mitchell, D., & Maloney, S. K. (2010). Physiological mechanisms in coping with climate change. *Physiological and Biochemical Zoology, 83*(5), 713–720. https://doi.org/10.1086/652242 (accessed on 20th February 2026).

Gaedke, U., Ruhenstroth-Bauer, M., Wiegand, I., Tirok, K., Aberle, N., Breithaupt, P., et al. (2010). Biotic interactions may overrule direct climate effects on spring phytoplankton dynamics. *Global Change Biology, 16*(3), 1122–1136. https://doi.org/10.1111/j.1365-2486.2009.02009.x (accessed on 20th February 2026).

Galbraith, E. D., Carozza, D. A., & Bianchi, D. (2017). A coupled human–Earth model perspective on long-term trends in the global marine fishery. *Nature Communications, 8*, 14884. https://doi.org/10.1038/ncomms14884 (accessed on 20th February 2026).

García Molinos, J., Halpern, B. S., Schoeman, D. S., Brown, C. J., Kiessling, W., Moore, P. J., et al. (2016). Climate velocity and the future global redistribution of marine biodiversity. *Nature Climate Change, 6*(1), 83–88. https://doi.org/10.1038/nclimate2769 (accessed on 20th February 2026).

García, R., & Ohnstone, R. W. (2006). Effects of *Lyngbya majuscula* (Cyanophycean) blooms on sediment nutrients and meiofaunal assemblages in seagrass beds in Moreton Bay, Australia. *Marine and Freshwater Research, 57*(2), 155–165. https://doi.org/10.1071/MF05053 (accessed on 20th February 2026).

Garzke, J., Ismar, S. M. H., & Sommer, U. (2015). Climate change affects low trophic level marine consumers: Warming decreases copepod size and abundance. *Ecologies, 177*(3), 849–860. https://doi.org/10.1007/s00442-014-3130-4 (accessed on 20th February 2026).

Gentili, F. G., & Fick, J. (2017). Algal cultivation in urban wastewater: An efficient way to reduce pharmaceutical pollutants. *Journal of Applied Phycology, 29*(1), 255–262. https://doi.org/10.1007/s10811-016-0950-0 (accessed on 20th February 2026).

Gigova, L., Ivanova, N., Gacheva, G., Andreeva, R., & Furnadzhieva, S. (2012). Response of *Trachydiscus minutus* (Xanthophyceae) to temperature and light. *Journal of Phycology, 48*(1), 85–93. https://doi.org/10.1111/j.1529-8817.2011.01088.x (accessed on 20th February 2026).

Gimenez, P. G. (2018). Climate conditions, and changes, affect microalgae communities… Should we worry? *Integrated Environmental Assessment and Management, 14*(2), 181–184. https://doi.org/10.1002/ieam.2009 (accessed on 20th February 2026).

Giordano, M., Beardall, J., & Raven, J. A. (2005). CO_2 concentrating mechanisms in algae: Mechanisms, environmental modulation, and evolution. *Annual Review of Plant Biology, 56*(1), 99–131. https://doi.org/10.1146/annurev.arplant.56.032604.144052 (accessed on 20th February 2026).

Gobler, C. J., & Sunda, W. G. (2012). Ecosystem disruptive algal blooms of the brown tide species, *Aureococcus anophagefferens* and *Aureoumbra lagunensis*. *Harmful Algae, 14*, 36–45. https://doi.org/10.1016/j.hal.2011.10.013 (accessed on 20th February 2026).

Gobler, C. J., Berry, D. L., Dyhrman, S. T., & Grigoriev, I. V. (2011). Niche of harmful alga *Aureococcus anophagefferens* revealed through economics. *Proceedings of the National Academy of Sciences of the United States of America, 108*(11), 4352–4357. https://doi.org/10.1073/pnas.1016106108 (accessed on 20th February 2026).

Gobler, C. J., Doherty, O. M., Hattenrath-Lehmann, T. K., Griffith, A. W., Kang, Y., & Litaker, R. W. (2017). Ocean warming since 1982 has expanded the niche of toxic algal blooms in the North Atlantic and North Pacific oceans. *Proceedings of the National Academy of Sciences of the United States of America, 114*(19), 4975–4980. https://doi.org/10.1073/pnas.1619575114 (accessed on 20th February 2026).

Goldman, J. C., & Ryther, J. H. (1976). Temperature-influenced species competition in mass cultures of marine phytoplankton. *Biotechnology and Bioengineering, 18*(8), 1125–1144. https://doi.org/10.1002/bit.260180809 (accessed on 20th February 2026).

Gonçalves, A. L., Pires, J. C. M., & Simões, M. (2016). *RSC Advances. Royal Society of Chemistry*, 1–31.

Granéli, E., Olsson, P., Carlsson, P., Granéli, W., & Christer, N. (1993). Weak "top-down" control of dinoflagellate growth in the coastal Skagerrak. *Journal of Plankton Research, 15*(2), 213–237. https://doi.org/10.1093/plankt/15.2.213 (accessed on 20th February 2026).

Green, M. A., Waldbusser, G. G., Reilly, S. L., Emerson, K., & O'Donnell, S. (2009). Death by dissolution: Sediment saturation state as a mortality factor for juvenile bivalves. *Limnology and Oceanography, 54*(4), 1037–1047. https://doi.org/10.4319/lo.2009.54.4.1037 (accessed on 20th February 2026).

Guerrini, F., Ciminiello, P., Dell'Aversano, C., Tartaglione, L., Fattorusso, E., Boni, L., et al. (2007). Influence of temperature, salinity and nutrient limitation on yessotoxin production and release by the dinoflagellate *Protoceratium reticulatum* in batch cultures. *Harmful Algae, 6*(5), 707–717. https://doi.org/10.1016/j.hal.2007.02.006 (accessed on 20th February 2026).

Guinotte, J. M., & Fabry, V. J. (2008). Ocean acidification and its potential effects on marine ecosystems. *Annals of the New York Academy of Sciences, 1134*, 320–342. https://doi.org/10.1196/annals.1439.013 (accessed on 20th February 2026).

Guo, Y., Yuan, Z., Xu, J., Wang, Z., Yuan, T., Zhou, W., et al. (2017). Metabolic acclimation mechanism in microalgae developed for CO_2 capture from industrial flue gas. *Algal Research, 26*, 225–233. https://doi.org/10.1016/j.algal.2017.07.029 (accessed on 20th February 2026).

Haley, S. T., Alexander, H., Juhl, A. R., & Dyhrman, S. T. (2017). Transcriptional response of the harmful raphidophyte *Heterosigma akashiwo* to nitrate and phosphate stress. *Harmful Algae, 68*, 258–270. https://doi.org/10.1016/j.hal.2017.07.001 (accessed on 20th February 2026).

Hallegraeff, G. M. (1993). Harmful algal blooms in the Australian region. *Marine Pollution Bulletin, 25*(5–8), 186–190. https://doi.org/10.1016/0025-326X(92)90223-S (accessed on 20th February 2026).

Hallegraeff, G. M. (2010). Ocean climate change, phytoplankton community responses, and harmful algal blooms: A formidable predictive challenge. *Journal of Phycology, 46*(2), 220–235. https://doi.org/10.1111/j.1529-8817.2010.00815.x (accessed on 20th February 2026).

Hansen, P. J. (2002). Effect of high pH on the growth and survival of marine phytoplankton: Implications for species succession. *Aquatic Microbial Ecology, 28*, 279–288.

Harke, M. J., Juhl, A. R., Haley, S. T., Alexander, H., & Dyhrman, S. T. (2017). Conserved transcriptional responses to nutrient stress in bloom-forming algae. *Frontiers in Microbiology, 8*, 1279. https://doi.org/10.3389/fmicb.2017.01279 (accessed on 20th February 2026).

Harke, M. J., Steffen, M. M., Gobler, C. J., Otten, T. G., Wilhelm, S. W., Wood, S. A., et al. (2016). A review of the global ecology, genomics, and biogeography of the toxic cyanobacterium, *Microcystis* spp. *Harmful Algae, 54*, 4–20. https://doi.org/10.1016/j.hal.2015.12.007 (accessed on 20th February 2026).

Hauck, J., & Völker, C. (2015). Rising atmospheric CO_2 leads to large impact of biology on Southern Ocean CO_2 uptake via changes of the Revelle factor. *Geophysical Research Letters, 42*(5), 1459–1464. https://doi.org/10.1002/2015GL063070 (accessed on 20th February 2026).

Heisler, J., Glibert, P. M., Burkholder, J. M., Anderson, D. M., Cochlan, W., Dennison, W. C., et al. (2008). Eutrophication and harmful algal blooms: A scientific consensus. *Harmful Algae, 8*(1), 3–13. https://doi.org/10.1016/j.hal.2008.08.006 (accessed on 20th February 2026).

Helbling, E. W., Buma, A. G. J., De Boer, M. K., & Villafañe, V. E. (2001). In situ impact of solar ultraviolet radiation on photosynthesis and DNA in temperate marine phytoplankton. *Marine Ecology Progress Series, 211*, 43–49. https://doi.org/10.3354/meps211043 (accessed on 20th February 2026).

Helbling, E. W., Villafañe, V. E., & Holm-Hansen, O. (1994). Effects of ultraviolet radiation on Antarctic marine phytoplankton photosynthesis with particular attention to the influence of mixing. *Antarctic Research Series, 62*, 207–227. https://doi.org/10.1029/AR062p0207 (accessed on 20th February 2026).

Hennon, G. M. M., & Dyhrman, S. T. (2020). Progress and promise of omics for predicting the impacts of climate change on harmful algal blooms. *Harmful Algae, 91*, 101587. https://doi.org/10.1016/j.hal.2019.03.005 (accessed on 20th February 2026).

Hennon, G. M. M., Williamson, O. M., Hernández Limón, M. D., Haley, S. T., & Dyhrman, S. T. (2019). Non-linear physiology and gene expression responses of harmful alga *Heterosigma akashiwo* to rising CO_2. *Protist, 170*(1), 38–51. https://doi.org/10.1016/j.protis.2018.10.002 (accessed on 20th February 2026).

Hessen, D. O., Van Donk, E., & Andersen, T. (1995). Growth responses, P-uptake and loss of flagella in *Chlamydomonas reinhardtii* exposed to UV-B. *Journal of Plankton Research, 17*(1), 17–27. https://doi.org/10.1093/plankt/17.1.17 (accessed on 20th February 2026).

Hirooka, S., Hirose, Y., Kanesaki, Y., & Miyagishima, S. y. (2017). Acidophilic green algal genome provides insights into adaptation to an acidic environment. *Proceedings of the National Academy of Sciences of the United States of America, 114*(39), E8304–E8313. https://doi.org/10.1073/pnas.1707072114 (accessed on 20th February 2026).

Hixson, S. M., & Arts, M. T. (2016). Climate warming is predicted to reduce omega-3 long-chain polyunsaturated fatty acid production in phytoplankton. *Global Change Biology, 22*(8), 2744–2755. https://doi.org/10.1111/gcb.13295 (accessed on 20th February 2026).

Holm-Hansen, O., Helbling, E. W., & Lubin, D. (1993). Ultraviolet radiation in Antarctica: Inhibition of primary production. *Photochemistry and Photobiology, 58*(4), 567–570. https://doi.org/10.1111/j.1751-1097.1993.tb04933.x (accessed on 20th February 2026).

Howes, E. L., Joos, F., Eakin, C. M., & Gattuso, J. P. (2015). An updated synthesis of the observed and projected impacts of climate change on the chemical, physical and biological processes in the oceans. *Frontiers in Marine Science, 2*, 36. https://doi.org/10.3389/fmars.2015.00036 (accessed on 20th February 2026).

Hussein, M. H., Abdullah, A. M., El Din, N. I. B., & Mishaqa, E. S. I. (2017). Biosorption potential of the microchlorophyte *Chlorella vulgaris* for some pesticides. *Journal of Fertilizers & Pesticides, 8*(1). https://doi.org/10.4172/2471-2728.1000177 (accessed on 20th February 2026).

Ibrahim, W. M., Karam, M. A., El-Shahat, R. M., & Adway, A. A. (2014). Biodegradation and utilization of organophosphorus pesticide malathion by cyanobacteria. *BioMed Research International, 2014*, 392682. https://doi.org/10.1155/2014/392682 (accessed on 20th February 2026).

Illoul, H., Rodríguez Hernández, F., Vila, M., Adjas, N., Aït Younes, A., Bournissa, M., et al. (2012). The genus *Ostreopsis* along the Algerian coastal waters (SW Mediterranean Sea) associated with a human respiratory intoxication episode. *Cryptogamie, Algologie, 33*(2), 209–216. https://doi.org/10.7872/crya.v33.iss2.2011.209 (accessed on 20th February 2026).

Intergovernmental Panel on Climate Change (IPCC). (2007). *Climate Change 2007: The Physical Science Basis. Contribution of Working Group I to the Fourth Assessment Report of the IPCC.* Cambridge University Press.

Intergovernmental Panel on Climate Change (IPCC). (2013). *Climate Change 2013: The Physical Science Basis. Contribution of Working Group I to the Fifth Assessment Report of the IPCC.* Cambridge University Press. https://doi.org/10.1017/CBO9781107415324

Intergovernmental Panel on Climate Change (IPCC). (2014). *Climate Change 2014: Impacts, Adaptation, and Vulnerability. Contribution of Working Group II to the Fifth Assessment Report of the IPCC.* Cambridge University Press.

Ivanina, A. V., & Sokolova, I. M. (2015). Interactive effects of metal pollution and ocean acidification on physiology of marine organisms. *Current Zoology, 61*(4), 653–668. https://doi.org/10.1093/czoolo/61.4.653 (accessed on 20th February 2026).

Jacques, G. (1983). Some Eco physiological aspects of the Antarctic phytoplankton. *Polar Biology, 2*(1), 27–33. https://doi.org/10.1007/BF00258282 (accessed on 20th February 2026).

Jaeckisch, N., Yang, I., Wohlrab, S., et al. (2011). Comparative genomic and transcriptomic characterization of the toxigenic marine dinoflagellate *Alexandrium ostenfeldii. PLoS ONE, 6*(12), e28012. https://doi.org/10.1371/journal.pone.0028012 (accessed on 20th February 2026).

Jia, J., Han, D., Gerken, H. G., et al. (2015). Molecular mechanisms for photosynthetic carbon partitioning into storage neutral lipids in *Nannochloropsis oceanica* under nitrogen-depletion conditions. *Algal Research, 7*, 66–77. https://doi.org/10.1016/j.algal.2014.11.005 (accessed on 20th February 2026).

Jiang, Y., Peng, X., Zhang, W., & Liu, T. (2012). Enhancement of acid resistance of *Scenedesmus dimorphus* by acid adaptation. *Journal of Applied Phycology, 24*(6), 1637–1641. https://doi.org/10.1007/s10811-012-9827-z (accessed on 20th February 2026).

Jin, Z. P., Luo, K., Zhang, S., Zheng, Q., & Yang, H. (2012). Bioaccumulation and catabolism of pyrometry in green algae. *Chemosphere, 87*(3), 278–284. https://doi.org/10.1016/j.chemosphere.2011.12.071 (accessed on 20th February 2026).

John, U., Beszteri, S., Glöckner, G., Singh, R., Medlin, L., & Cembella, A. D. (2010). Genomic characterization of the ichthyotoxic prymnesiophyte *Chrysochromulina polylepis*, and the expression of polyketide synthase genes in synchronized cultures. *European Journal of Phycology, 45*(3), 215–229. https://doi.org/10.1080/09670261003746193 (accessed on 20th February 2026).

Johnson, V. R., Brownlee, C., Rickaby, R. E. M., Graziano, M., Milazzo, M., & Hall-Spencer, J. M. (2013). Responses of marine benthic microalgae to elevated CO_2. *Marine*

Biology, 160(8), 1813–1824. https://doi.org/10.1007/s00227-011-1840-2 (accessed on 20th February 2026).

Kanekar, P., Bhadbhade, B., Deshpande, N., & Sarnaik, S. (2004). Biodegradation of organophosphorus pesticides. *Proceedings of the Indian National Science Academy, B70*(1), 57–70.

Kang, C. K., Kim, J. B., Lee, K. S., Lee, P. Y., & Hong, J. S. (2003). Trophic importance of benthic microalgae to macro zoobenthos in coastal bay systems in Korea: Dual stable C and N isotope analyses. *Marine Ecology Progress Series, 259*, 79–92. https://doi.org/10.3354/meps259079 (accessed on 20th February 2026).

Karentz, D., Cleaver, J. E., & Mitchell, D. L. (1991). Cell survival characteristics and molecular responses of Antarctic phytoplankton to ultraviolet-B radiation. *Journal of Phycology, 27*, 326–341.

Keeling, P. J., Burki, F., Wilcox, H. M., Allam, B., Allen, E. E., Amaral-Zettler, L. A., et al. (2014). The Marine Microbial Eukaryote Transcriptome Sequencing Project (MMETSP): Illuminating the functional diversity of eukaryotic life in the oceans through transcriptome sequencing. *PLoS Biology, 12*(6), e1001889. https://doi.org/10.1371/journal.pbio.1001889 (accessed on 20th February 2026).

Kemp, W. M., Boynton, W. R., Adolf, J. E., Boesch, D. F., Boicourt, W. C., Brush, G., et al. (2005). Eutrophication of Chesapeake Bay: Historical trends and ecological interactions. *Marine Ecology Progress Series, 303*, 1–29. https://doi.org/10.3354/meps303001 (accessed on 20th February 2026).

Khalil, Z. I., Asker, M. M. S., El-Sayed, S., & Kobbia, I. A. (2010). Effect of pH on growth and biochemical responses of *Dunaliella bardawil* and *Chlorella ellipsoidea*. *World Journal of Microbiology and Biotechnology, 26*(7), 1225–1231. https://doi.org/10.1007/s11274-009-0292-z (accessed on 20th February 2026).

Kleypas, J. A., Feely, R. A., Fabry, V. J., Langdon, C., Sabine, C. L., & Robbins, L. L. (2006). Impacts of ocean acidification on coral reefs and other marine calcifiers: A guide for future research. *Atmospheric Research, 18*(1), 1–85.

Knutzen, J. (1981). Effects of decreased pH on marine organisms. *Marine Pollution Bulletin, 12*(1), 25–29. https://doi.org/10.1016/0025-326X(81)90136-3 (accessed on 20th February 2026).

Ko, G. W. K., Dineshram, R., Campanati, C., Chan, V. B. S., Havenhand, J., & Thiyagarajan, V. (2014). Interactive effects of ocean acidification, elevated temperature, and reduced salinity on early-life stages of the Pacific oyster. *Environmental Science & Technology, 48*(17), 10079–10088. https://doi.org/10.1021/es501611u (accessed on 20th February 2026).

Kremp, A., Godhe, A., Egardt, J., Dupont, S., Suikkanen, S., Casabianca, S., et al. (2012). Intraspecific variability in the response of bloom-forming marine microalgae to changed climate conditions. *Ecology and Evolution, 2*(6), 1195–1207. https://doi.org/10.1002/ece3.245 (accessed on 20th February 2026).

Kroeker, K. J., Kordas, R. L., Crim, R. N., & Singh, G. G. (2010). Meta-analysis reveals negative yet variable effects of ocean acidification on marine organisms. *Ecology Letters, 13*(11), 1419–1434. https://doi.org/10.1111/j.1461-0248.2010.01518.x (accessed on 20th February 2026).

Kuffner, I. B., Andersson, A. J., Jokiel, P. L., Rodgers, K. S., & MacKenzie, F. T. (2008). Decreased abundance of crustose coralline algae due to ocean acidification. *Nature*

Geoscience, 1(2), 114–117. https://doi.org/10.1038/ngeo100 (accessed on 20th February 2026).

Kumar, K., Dasgupta, C. N., Nayak, B., Lindblad, P., & Das, D. (2011). Development of suitable photobioreactors for CO_2 sequestration addressing global warming using green algae and cyanobacteria. *Bioresource Technology, 102*(8), 4945–4953. https://doi.org/10.1016/j.biortech.2011.01.054 (accessed on 20th February 2026).

Kurade, M. B., Kim, J. R., Govindwar, S. P., & Jeon, B. H. (2016). Insights into microalgae-mediated biodegradation of diazinon by *Chlorella vulgaris*: Microalgal tolerance to xenobiotic pollutants and metabolism. *Algal Research, 20*, 126–134. https://doi.org/10.1016/j.algal.2016.10.003 (accessed on 20th February 2026).

Kwiatkowski, L., Aumont, O., & Bopp, L. (2019). Consistent trophic amplification of marine biomass declines under climate change. *Global Change Biology, 25*(1), 218–229. https://doi.org/10.1111/gcb.14468 (accessed on 20th February 2026).

Kwiatkowski, L., Aumont, O., Bopp, L., & Ciais, P. (2018). The impact of variable phytoplankton stoichiometry on projections of primary production, food quality, and carbon uptake in the global ocean. *Global Biogeochemical Cycles, 32*(4), 516–528. https://doi.org/10.1002/2017GB005799 (accessed on 20th February 2026).

Le Faucheur, S., Campbell, P. G. C., Fortin, C., & Slaveykova, V. I. (2014). Interactions between mercury and phytoplankton: Speciation, bioavailability, and internal handling. *Environmental Toxicology and Chemistry, 33*(6), 1211–1224. https://doi.org/10.1002/etc.2424 (accessed on 20th February 2026).

Lefort, S., Aumont, O., Bopp, L., Arsouze, T., Gehlen, M., & Maury, O. (2015). Spatial and body-size dependent response of marine pelagic communities to project global climate change. *Global Change Biology, 21*(1), 154–164. https://doi.org/10.1111/gcb.12679 (accessed on 20th February 2026).

Leng, L., Wei, L., Xiong, Q., Xu, S., Li, W., Lv, S., et al. (2020). Use of microalgae-based technology for the removal of antibiotics from wastewater: A review. *Chemosphere, 238*, 124680.

Leong, Y. K., & Chang, J. S. (2020). Bioremediation of heavy metals using microalgae: Recent advances and mechanisms. *Bioresource Technology*, 122886. https://doi.org/10.1016/j.biortech.2020.122886 (accessed on 20th February 2026).

Letterio, G., Minutoli, R., Bergamasco, A., Granata, A., Zagami, G., & Antezana, T. (2011). Short-term changes in zooplankton community in Paso Ancho Basin (Strait of Magellan): Functional trophic structure and diel vertical migration. *Polar Biology, 34*(9), 1301–1317. https://doi.org/10.1007/s00300-011-1031-0 (accessed on 20th February 2026).

Lewis, R. J., & Ruff, T. A. (1993). Ciguatera: Ecological, clinical, and socioeconomic perspectives. *Critical Reviews in Environmental Science and Technology, 23*(2), 137–156. https://doi.org/10.1080/10643389309388447 (accessed on 20th February 2026).

Li, W. K. W. (1980). Temperature adaptation in phytoplankton: Cellular and photosynthetic characteristics. In P. G. Falkowski (Ed.), *Primary Productivity in the Sea* (pp. 259–279). Springer. https://doi.org/10.1007/978-1-4684-3890-1 (accessed on 20th February 2026).

Li, W. K. W. (1985). Photosynthetic response to temperature of marine phytoplankton along a latitudinal gradient (16°N to 74°N). *Deep Sea Research Part A: Oceanographic*

Research Papers, 32(11), 1385–1393. https://doi.org/10.1016/0198-0149(85)90054-8 (accessed on 20th February 2026).

Li, W. K. W., Smith, J. C., & Platt, T. (1984). Temperature response of photosynthetic capacity and carboxylase activity in Arctic marine phytoplankton. *Marine Ecology Progress Series, 17*, 237–243. https://doi.org/10.3354/meps017237 (accessed on 20th February 2026).

Li, W., Xu, X., Fujibayashi, M., Niu, Q., Tanaka, N., & Nishimura, O. (2016). Response of microalgae to elevated CO_2 and temperature: Impact of climate change on freshwater ecosystems. *Environmental Science and Pollution Research, 23*(19), 19847–19860. https://doi.org/10.1007/s11356-016-7180-5 (accessed on 20th February 2026).

Lindberg, R. T., & Collins, S. (2020). Quality–quantity trade-offs drive functional trait evolution in a model microalgal "climate change winner". *Ecology Letters, 23*(5), 780–790. https://doi.org/10.1111/ele.13478 (accessed on 20th February 2026).

Litchman, E., & Neale, P. J. (2005). UV effects on photosynthesis, growth and acclimation of an estuarine diatom and cryptomonad. *Marine Ecology Progress Series, 300*, 53–62. https://doi.org/10.3354/meps300053 (accessed on 20th February 2026).

Liu, Z., Campbell, V., Heidelberg, K. B., & Caron, D. A. (2016). Gene expression characterizes different nutritional strategies among three mixotrophic protists. *FEMS Microbiology Ecology, 92*(7), fiw106. https://doi.org/10.1093/femsec/fiw106 (accessed on 20th February 2026).

Lotze, H. K., Tittensor, D. P., Bryndum-Buchholz, A., & Worm, B. (2019). Global ensemble projections reveal trophic amplification of ocean biomass declines with climate change. *Proceedings of the National Academy of Sciences of the United States of America, 116*(26), 12907–12912. https://doi.org/10.1073/pnas.1900194116 (accessed on 20th February 2026).

Lundholm, N., Skov, J., Pocklington, R., & Moestrup, Ø. (1994). Domoic acid, the toxic amino acid responsible for amnesic shellfish poisoning, now in *Pseudonitzschia seriata* (Bacillariophyceae) in Europe. *Phycological, 33*(6), 475–478. https://doi.org/10.1080/0013191640160307 (accessed on 20th February 2026).

Lynch, P. (2008). The origins of computer weather prediction and climate modeling. *Journal of Computational Physics, 227*(7), 3431–3444. https://doi.org/10.1016/j.jcp.2007.02.034 (accessed on 20th February 2026).

MacIntyre, H. L., Kana, T. M., Anning, T., & Geider, R. J. (2002). Photo acclimation of photosynthesis–irradiance response curves and photosynthetic pigments in microalgae and cyanobacteria. *Journal of Phycology, 38*(1), 17–38.

Maestrini, S. Y., & Granéli, E. (1991). Environmental conditions and Eco physiological mechanisms which led to the 1988 *Chrysochromulina polylepis* bloom: A hypothesis. *Oceanological Acta, 14*(4), 397–413.

Magdaong, N. C. M., & Blankenship, R. E. (2018). Photoprotective, excited-state quenching mechanisms in diverse photosynthetic organisms. *Journal of Biological Chemistry, 293*(14), 5018–5025. https://doi.org/10.1074/jbc.TM117.000233 (accessed on 20th February 2026).

Makower, A. K., Schuurmans, J. M., Groth, D., Zilliges, Y., Matthijs, H. C. P., & Dittmann, E. (2015). Transcriptomics-aided dissection of the intracellular and extracellular roles of microcystin in *Microcystis aeruginosa* PCC 7806. *Applied and Environmental*

Microbiology, 81(2), 544–554. https://doi.org/10.1128/AEM.02601-14 (accessed on 20th February 2026).

Mangoni, O., Saggiomo, V., Bolinesi, F., Margiotta, F., Budillon, G., Cotroneo, Y., et al. (2017). Phytoplankton blooms during austral summer in the Ross Sea, Antarctica: Driving factors and trophic implications. *PLoS ONE, 12*(4), e0176033. https://doi.org/10.1371/journal.pone.0176033 (accessed on 20th February 2026).

Marchello, A. E., Lombardi, A. T., Dellamano-Oliveira, M. J., & De Souza, C. O. W. (2015). Microalgae population dynamics in photobioreactors with secondary sewage effluent as culture medium. *Brazilian Journal of Microbiology, 46*(1), 75–84. https://doi.org/10.1590/S1517-838246120131225 (accessed on 20th February 2026).

Masó, M., & Garcés, E. (2006). Harmful microalgae blooms (HAB): Problematic and conditions that induce them. *Marine Pollution Bulletin, 53*(10–12), 620–630. https://doi.org/10.1016/j.marpolbul.2006.08.006 (accessed on 20th February 2026).

Mata, T. M., Martins, A. A., & Caetano, N. S. (2010). Microalgae for biodiesel production and other applications: A review. *Renewable and Sustainable Energy Reviews, 14*(1), 217–232. https://doi.org/10.1016/j.rser.2009.07.020 (accessed on 20th February 2026).

Matamoros, V., Uggetti, E., García, J., & Bayona, J. M. (2016). Assessment of the mechanisms involved in the removal of emerging contaminants by microalgae from wastewater: A laboratory-scale study. *Journal of Hazardous Materials, 301*, 197–205. https://doi.org/10.1016/j.jhazmat.2015.08.050 (accessed on 20th February 2026).

Matozzo, V., Chinellato, A., Munari, M., Finos, L., Bressan, M., & Marin, M. G. (2012). First evidence of immunomodulation in bivalves under seawater acidification and increased temperature. *PLoS ONE, 7*(3), e33820. https://doi.org/10.1371/journal.pone.0033820 (accessed on 20th February 2026).

Matsunaga, T., Takeyama, H., Nakao, T., & Yamazawa, A. (1999). Screening of marine microalgae for bioremediation of cadmium-polluted seawater. *Progress in Industrial Microbiology, 35*, 33–38. https://doi.org/10.1016/S0079-6352(99)80095-2 (accessed on 20th February 2026).

McEwan, M., Humayun, R., Slamovits, C. H., & Keeling, P. J. (2008). Nuclear genome sequence survey of the dinoflagellate *Heterocapsa triquetra. Journal of Eukaryotic Microbiology, 55*(6), 530–535. https://doi.org/10.1111/j.1550-7408.2008.00357.x (accessed on 20th February 2026).

Meguro, H., Ito, K., & Fukushima, H. (1967). Ice flora (bottom type): A mechanism of primary production in polar seas and the growth of diatoms in sea ice. *Arctic, 20*(2), 114–133. https://doi.org/10.14430/arctic3287 (accessed on 20th February 2026).

Menció, A., Mas-Pla, J., Otero, N., Regàs, O., Boy-Roura, M., Puig, R., et al. (2016). Nitrate pollution of groundwater; all right… but nothing else? *Science of the Total Environment, 539*, 241–251. https://doi.org/10.1016/j.scitotenv.2015.08.151 (accessed on 20th February 2026).

Michel, C., Legendre, L., Therriault, J. C., & Demers, S. (1989). Photosynthetic responses of Arctic sea-ice microalgae to short-term temperature acclimation. *Polar Biology, 9*(7), 437–442. https://doi.org/10.1007/BF00443230 (accessed on 20th February 2026).

Miller, A. W., Reynolds, A. C., Sobrino, C., & Riedel, G. F. (2009). Shellfish face uncertain future in high CO_2 world: Influence of acidification on oyster larvae calcification and growth in estuaries. *PLoS ONE, 4*(5), e5661. https://doi.org/10.1371/journal.pone.0005661 (accessed on 20th February 2026).

Miraglia, M., Marvin, H. J. P., Kleter, G. A., Battilani, P., Brera, C., Coni, E., et al. (2009). Climate change and food safety: An emerging issue with special focus on Europe. *Food and Chemical Toxicology, 47*(5), 1009–1021. https://doi.org/10.1016/j.fct.2009.02.005

Mishra, A., Medhi, K., Malaviya, P., & Thakur, I. S. (2019). Omics approaches for microalgal applications: Prospects and challenges. *Bioresource Technology, 291*, 121890. https://doi.org/10.1016/j.biortech.2019.121890 (accessed on 20th February 2026).

Mitra, A., & Flynn, K. J. (2005). Predator–prey interactions: Is "ecological stoichiometry" sufficient when good food goes bad? *Journal of Plankton Research, 27*(5), 393–399. https://doi.org/10.1093/plankt/fbi022 (accessed on 20th February 2026).

Molina-Santos, M., Terneus-Jácome, E., Yánez-Moretta, P., & Cueva-Sánchez, M. (2018). Resiliencia de la comunidad fitoplanctónica en la laguna andina de Papallacta y sus afluentes, ocho años después de un derrame petrolero. *28*(2), 67–83.

Moore, S. K., Mantua, N. J., Hickey, B. M., & Trainer, V. L. (2009). Recent trends in paralytic shellfish toxins in Puget Sound: Relationships to climate, and capacity for prediction of toxic events. *Harmful Algae, 8*(3), 463–477. https://doi.org/10.1016/j.hal.2008.10.003 (accessed on 20th February 2026).

Morey, J. S., Monroe, E. A., Kinney, A. L., Beal, M., Johnson, J. G., Hitchcock, G. L., et al. (2011). Transcriptomic response of the red tide dinoflagellate, *Karenia brevis*, to nitrogen and phosphorus depletion and addition. *BMC Genomics, 12*, 346. https://doi.org/10.1186/1471-2164-12-346 (accessed on 20th February 2026).

Moss, B., McKee, D., Atkinson, D., Collings, S. E., Eaton, J. W., Gill, A. B., et al. (2003). How important is climate? Effects of warming, nutrient addition and fish on phytoplankton in shallow lake microcosms. *Journal of Applied Ecology, 40*, 782–792. https://doi.org/10.1046/j.1365-2664.2003.00839.x (accessed on 20th February 2026).

Muñoz, J., & Merrett, M. J. (1989). Inorganic carbon transport in some marine eukaryotic microalgae. *Planta, 178*(4), 450–455. https://doi.org/10.1007/BF00963814 (accessed on 20th February 2026).

Murphy, G. E. P., Romanuk, T. N., & Worm, B. (2020). Cascading effects of climate change on plankton community structure. *Ecology and Evolution, 10*, 2170–2181. https://doi.org/10.1002/ece3.6055 (accessed on 20th February 2026).

Murray, R. W., Leinen, M., Murray, D. W., Mix, A. C., & Knowlton, C. W. (1995). Terrigenous Fe input and biogenic sedimentation in the glacial and interglacial equatorial Pacific Ocean. *Global Biogeochemical Cycles, 9*(4), 667–684. https://doi.org/10.1029/95GB02833 (accessed on 20th February 2026).

Murray, S. A., Garby, T., Hoppenrath, M., & Neilan, B. A. (2012). Genetic diversity, morphological uniformity and polyketide production in dinoflagellates (*Amphidinium*, Dinoflagellata). *PLoS ONE, 7*(6), e38253. https://doi.org/10.1371/journal.pone.0038253 (accessed on 20th February 2026).

Nalewajko, C., Colman, B., & Olaveson, M. (1997). Effects of pH on growth, photosynthesis, respiration, and copper tolerance of three *Scenedesmus* strains. *Environmental and Experimental Botany, 37*(2, 3), 153–160. https://doi.org/10.1016/S0098-8472(96)01029-5 (accessed on 20th February 2026).

Neale, P. J., Banaszak, A. T., & Jarriel, C. R. (1998). Ultraviolet sunscreens in *Gymnodinium sanguineum* (Dinophyceae): Mycosporine-like amino acids protect against inhibition of photosynthesis. *Journal of Phycology, 34*(6), 928–938. https://doi.org/10.1046/j.1529-8817.1998.340928.x (accessed on 20th February 2026).

Neori, A., & Holm-Hansen, O. (1982). Effect of temperature on rate of photosynthesis in Antarctic phytoplankton. *Polar Biology, 1*(1), 33–38. https://doi.org/10.1007/BF00568752 (accessed on 20th February 2026).

Nie, J., Sun, Y., Zhou, Y., Kumar, M., Usman, M., Li, J., et al. (2020). Bioremediation of water containing pesticides by microalgae: Mechanisms, methods, and prospects for future research. *Science of the Total Environment, 707*, 136080. https://doi.org/10.1016/j.scitotenv.2019.136080 (accessed on 20th February 2026).

Nielsen, E. S., & Hansen, V. K. (1959). Light adaptation in marine phytoplankton populations and its interrelation with temperature. *Physiologiae Plantarum, 12*(2), 353–370. https://doi.org/10.1111/j.1399-3054.1959.tb07939.x (accessed on 20th February 2026).

Nimer, N. A., & Merrett, M. J. (1992). Utilization of inorganic carbon by the coccolithophorid *Emiliania huxleyi* (Lohmann) Kamptner. *New Phytologist, 121*(1), 173–177. https://doi.org/10.1111/j.1469-8137.1992.tb01068.x (accessed on 20th February 2026).

O'Connor, M. I., Piehler, M. F., Leech, D. M., Anton, A., & Bruno, J. F. (2009). Warming and resource availability shift food web structure and metabolism. *PLoS Biology, 7*(8), e1000178. https://doi.org/10.1371/journal.pbio.1000178 (accessed on 20th February 2026).

Orr, J. C., Fabry, V. J., Aumont, O., Bopp, L., Doney, S. C., Feely, R. A., et al. (2005). Anthropogenic ocean acidification over the twenty-first century and its impact on calcifying organisms. *Nature, 437*(7059), 681–686. https://doi.org/10.1038/nature04095 (accessed on 20th February 2026).

Ortiz-Hernández, L., Sánchez-Salinas, E., Dantán-González, E., & Castrejón-Godínez, M. L. (2013). Pesticide biodegradation: Mechanisms, genetics and strategies to enhance the process. In *Biodegradation: Life of Science* (pp. 251–280).

Ota, M., Takenaka, M., Sato, Y., Smith, R. L., & Inomata, H. (2015). Effects of light intensity and temperature on photoautotrophic growth of a green microalga, *Chlorococcum littorale*. *Biotechnology Reports, 7*, 24–29. https://doi.org/10.1016/j.btre.2015.05.001 (accessed on 20th February 2026).

Palmisano, A. C., SooHoo, J. B., & Sullivan, C. W. (1987). Effects of four environmental variables on photosynthesis–irradiance relationships in Antarctic sea-ice microalgae. *Marine Biology, 94*(2), 299–306. https://doi.org/10.1007/BF00392944 (accessed on 20th February 2026).

Palmqvist, K. (2000). Carbon economy in lichens. *New Phytologist, 148*(1), 11–36. https://doi.org/10.1046/j.1469-8137.2000.00732.x (accessed on 20th February 2026).

Pardew, J., Pimentel, M. B., & Low-Décarie, E. (2018). Predictable ecological response to rising CO_2 of a community of marine phytoplankton. *Ecology and Evolution, 8*(8), 4292–4302. https://doi.org/10.1002/ece3.3971 (accessed on 20th February 2026).

Park, J. B. K., Craggs, R. J., & Shilton, A. N. (2011). Wastewater treatment high-rate algal ponds for biofuel production. *Bioresource Technology, 102*(1), 35–42. https://doi.org/10.1016/j.biortech.2010.06.158 (accessed on 20th February 2026).

Pawlowiez, R., Morey, J. S., Darius, H. T., Chinain, M., & Van Dolah, F. M. (2014). Transcriptome sequencing reveals single-domain type I-like polyketide synthases in the toxic dinoflagellate *Gambierdiscus polynesiensis*. *Harmful Algae, 36*, 29–37. https://doi.org/10.1016/j.hal.2014.04.013 (accessed on 20th February 2026).

Paz, B., Vázquez, J. A., Riobó, P., & Franco, J. M. (2006). Study of the effect of temperature, irradiance and salinity on growth and yessotoxin production by the dinoflagellate *Protoceratium reticulatum* in culture by using a kinetic and factorial approach. *Marine Environmental Research, 62*(4), 286–300. https://doi.org/10.1016/j.marenvres.2006.04.066 (accessed on 20th February 2026).

Pearson, L. A., Dittmann, E., Mazmouz, R., Ongley, S. E., D'Agostino, P. M., & Neilan, B. A. (2016). The genetics, biosynthesis and regulation of toxic specialized metabolites of cyanobacteria. *Harmful Algae, 54*, 98–111. https://doi.org/10.1016/j.hal.2015.11.002 (accessed on 20th February 2026).

Peletier, H., Gieskes, W. W. C., & Buma, A. G. J. (1996). Ultraviolet-B radiation resistance of benthic diatoms isolated from tidal flats in the Dutch Wadden Sea. *Marine Ecology Progress Series, 135*(1–3), 163–168. https://doi.org/10.3354/meps135163 (accessed on 20th February 2026).

Pereira, K. C., Costa, P. M., Costa, M. H., Luque, Á., DelValls, T. A., & Riba López, I. (2016). Effects of the increase of temperature and CO_2 concentration on the polychaete *Nereis diversicolor*: Simulating extreme scenarios of climate change in marine sediments. *Hydrobiological, 772*(1), 161–174. https://doi.org/10.1007/s10750-016-2656-3 (accessed on 20th February 2026).

Pérez-García, O., Escalante, F. M. E., de-Bashan, L. E., & Bashan, Y. (2011). Heterotrophic cultures of microalgae: Metabolism and potential products. *Water Research, 45*(1), 11–36. https://doi.org/10.1016/j.watres.2010.08.037

Pérez-Legaspi, I. A., Ortega-Clemente, L. A., Moha-León, J. D., Ríos-Leal, E., Curiel-Ramírez Gutiérrez, S., & Rubio-Franchini, I. (2016). Effect of the pesticide lindane on the biomass of the microalgae *Nannochloris oculata*. *Journal of Environmental Science and Health, Part B, 51*(2), 103–106. https://doi.org/10.1080/03601234.2015.1092824 (accessed on 20th February 2026).

Pham, T. L. (2020). Using benthic diatoms as a bioindicator to assess rural–urban river conditions in tropical area: A case study in the Sai Gon River, Vietnam. *Pollution, 6*(2), 387–398. https://doi.org/10.22059/POLL.2020.292996.716 (accessed on 20th February 2026).

Pinsky, M. L., Worm, B., Fogarty, M. J., Sarmiento, J. L., & Levin, S. A. (2013). Climate velocities. *Science, 341*(6151), 1239–1242.

Pomerleau, C., Stern, G. A., Pućko, M., Foster, K. L., Macdonald, R. W., & Fortier, L. (2016). Pan-Arctic concentrations of mercury and stable isotope ratios of carbon ($\delta^{13}C$) and nitrogen ($\delta^{15}N$) in marine zooplankton. *Science of the Total Environment, 551–552*, 92–100. https://doi.org/10.1016/j.scitotenv.2016.01.172 (accessed on 20th February 2026).

Ponmani, T., Guo, R., & Ki, J. S. (2016). Analysis of the genomic DNA of the harmful dinoflagellate *Prorocentrum minimum*: A brief survey focused on the noncoding RNA gene sequences. *Journal of Applied Phycology, 28*(1), 335–344. https://doi.org/10.1007/s10811-015-0570-0 (accessed on 20th February 2026).

Popin, R. V., Rigonato, J., Abreu, V. A. C., Andreote, A. P. D., Silveira, S. B., Odebrecht, C., et al. (2016). Draft genome assembly of the bloom-forming cyanobacterium *Nodularia spumigena* strain CENA596 in shrimp production ponds. *Genome Announcements, 4*(3), e00466-16. https://doi.org/10.1128/genomeA.00466-16 (accessed on 20th February 2026).

Pörtner, H. O. (2008). Ecosystem effects of ocean acidification in times of ocean warming: A physiologist's view. *Marine Ecology Progress Series, 373*, 203–217. https://doi.org/10.3354/meps07768 (accessed on 20th February 2026).

Pörtner, H. O., Karl, D. M., Boyd, P. W., et al. (2014). Ocean systems. In C. B. Field et al. (Eds.), *Climate Change 2014: Impacts, Adaptation, and Vulnerability. Part A: Global and Sectoral Aspects* (pp. 411–484). Cambridge University Press.

Poulin, R. X., Poulson-Ellestad, K. L., Roy, J. S., & Kubanek, J. (2018). Variable allelopathy among phytoplankton reflected in red tide metabolome. *Harmful Algae, 71*, 50–56. https://doi.org/10.1016/j.hal.2017.12.002 (accessed on 20th February 2026).

Poulson-Ellestad, K. L., Jones, C. M., Roy, J., & Nunn, B. L. (2014). Metabolomics and proteomics reveal impacts of chemically mediated competition on marine plankton. *Proceedings of the National Academy of Sciences of the United States of America, 111*(25), 9009–9014. https://doi.org/10.1073/pnas.1402130111 (accessed on 20th February 2026).

Prézelin, B. B., Boucher, N. P., & Smith, R. C. (1993). Response: Polyethylene bags and solar ultraviolet radiation. *Science, 259*(5094), 534–535.

Priyadarshani, I., Sahu, D., & Rath, B. (2012). Microalgal bioremediation: Current practices and perspectives. *Journal of Biochemical Technology, 3*(3), 299–304.

Ragg, N. L. C., King, N., Watts, E., & Morrish, J. (2010). Optimizing the delivery of the key dietary diatom *Chaetoceros calcitrans* to intensively cultured Green shell™ mussel larvae, *Perna canaliculus*. *Aquaculture, 306*(1–4), 270–280. https://doi.org/10.1016/j.aquaculture.2010.05.010 (accessed on 20th February 2026).

Ratti, S., Giordano, M., & Morse, D. (2007). CO_2-concentrating mechanisms of the potentially toxic dinoflagellate *Protoceratium reticulatum* (Dinophyceae, Gonyaulacales). *Journal of Phycology, 43*(4), 693–701. https://doi.org/10.1111/j.1529-8817.2007.00368.x (accessed on 20th February 2026).

Raven, J. A., & Geider, R. J. (1988). Temperature and algal growth. *New Phytologist, 110*(4), 441–461. https://doi.org/10.1111/j.1469-8137.1988.tb00282.x (accessed on 20th February 2026).

Raven, J. A., & Johnston, A. (1991). Mechanisms of inorganic-carbon acquisition in marine phytoplankton and their implications for the use of other resources. *Limnology and Oceanography, 36*(8), 1701–1714.

Raven, J. A., Kübler, J. E., & Beardall, J. (2000). Put out the light and then put out the light. *Journal of the Marine Biological Association of the United Kingdom, 80*(1), 1–25. https://doi.org/10.1017/S0025315499001526 (accessed on 20th February 2026).

Raven, J., Caldeira, K., Elderfield, H., et al. (2005). *Ocean Acidification Due to Increasing Atmospheric Carbon Dioxide*. The Royal Society.

Reid, G. K., Gurney-Smith, H. J., Flaherty, M., Garber, A. F., Forster, I., Brewer-Dalton, K., et al. (2019). Climate change and aquaculture: Considering adaptation potential. *Aquaculture Environment Interactions, 11*, 603–624. https://doi.org/10.3354/aei00333 (accessed on 20th February 2026).

Renberg, L., Johansson, A. I., Shutova, T., Stenlund, H., Aksmann, A., Raven, J. A., et al. (2010). A metabolomic approach to study major metabolite changes during acclimation to limiting CO_2 in *Chlamydomonas reinhardtii*. *Plant Physiology, 154*(1), 187–196. https://doi.org/10.1104/pp.110.157651 (accessed on 20th February 2026).

Reynolds, C. S. (1989). *Plankton Ecology*. Brock/Springer Series in Contemporary Bioscience. Springer.

Riebesell, U., Zondervan, I., Rost, B., Tortell, P. D., Zeebe, R. E., & Morel, F. M. M. (2000). Reduced calcification of marine plankton in response to increased atmospheric

CO_2. *Nature, 407*(6802), 364–367. https://doi.org/10.1038/35030078 (accessed on 20th February 2026).
Roberts, D. A., Birchenough, S. N. R., Lewis, C., Sanders, M. B., Bolam, T., & Sheahan, D. (2013). Ocean acidification increases the toxicity of contaminated sediments. *Global Change Biology, 19*(2), 340–351. https://doi.org/10.1111/gcb.12048 (accessed on 20th February 2026).
Rochet, M., Legendre, L., & Demers, S. (1985). Acclimation of sea-ice microalgae to freezing temperature. *Marine Ecology Progress Series, 24*, 187–191. https://doi.org/10.3354/meps024187 (accessed on 20th February 2026).
Rost, B., Richter, K. U., Riebesell, U., & Hansen, P. J. (2006). Inorganic carbon acquisition in red tide dinoflagellates. *Plant, Cell & Environment, 29*(5), 810–822. https://doi.org/10.1111/j.1365-3040.2005.01450.x (accessed on 20th February 2026).
Roux, R., Gosselin, M., Desrosiers, G., & Nozais, C. (2002). Effects of reduced UV radiation on a microbenthic community during a microcosm experiment. *Marine Ecology Progress Series, 225*, 29–43. https://doi.org/10.3354/meps225029 (accessed on 20th February 2026).
Sabine, C. L., Feely, R. A., Gruber, N., Key, R. M., Lee, K., Bullister, J. L., et al. (2004). The oceanic sink for anthropogenic CO_2. *Science, 305*(5682), 367–371.
Salisbury, J., Green, M., Hunt, C., & Campbell, J. (2008). Coastal acidification by rivers: A threat to shellfish? *Eos, 89*(50), 513. https://doi.org/10.1029/2008EO500001 (accessed on 20th February 2026).
Sandrini, G., Cunsolo, S., Schuurmans, J. M., Matthijs, H. C. P., & Huisman, J. (2015). Changes in gene expression, cell physiology and toxicity of the harmful cyanobacterium *Microcystis aeruginosa* at elevated CO_2. *Frontiers in Microbiology, 6*, 401. https://doi.org/10.3389/fmicb.2015.00401 (accessed on 20th February 2026).
Sassenhagen, I., Wilken, S., Godhe, A., & Rengefors, K. (2015). Phenotypic plasticity and differentiation in an invasive freshwater microalga. *Harmful Algae, 41*, 38–45. https://doi.org/10.1016/j.hal.2014.11.001 (accessed on 20th February 2026).
Savage, R. E. (1932). *Phaeocystis and herring shoals*. *Journal of Ecology, 20*(2), 326–340.
Schaum, C. (2018). Phosphorus removal in wastewater treatment plants. In *Phosphorus: Polluter and Resource of the Future–Removal and Recovery from Wastewater* (Vol. 17). IWA Publishing.
Schaum, C., Rost, B., Millar, A. J., & Collins, S. (2013). Variation in plastic responses of a globally distributed picoplankton species to ocean acidification. *Nature Climate Change, 3*(3), 298–302. https://doi.org/10.1038/nclimate1774 (accessed on 20th February 2026).
Schmidt, D., & Boyd, P. W. (2016). Forecast ocean variability. *Nature, 539*(7628), 162, 163. https://doi.org/10.1038/539162a (accessed on 20th February 2026).
Shelford, V. E. (2013). Some concepts of bioecology. *Ecology, 12*(3), 455–467.
Shumway, S. E., van Egmond, H. P., Hurst, J. W., & Bean, L. L. (1995). Management of shellfish resources. In *Manual on Harmful Marine Microalgae* (pp. 433–461).
Singh, A., Ummalyma, S. B., & Sahoo, D. (2020). Bioremediation and biomass production of microalgae cultivation in river water contaminated with pharmaceutical effluent. *Bioresource Technology, 307*, 123233.
Singh, S. P., & Singh, P. (2015). Effect of temperature and light on the growth of algae species: A review. *Renewable and Sustainable Energy Reviews, 50*, 431–444. https://doi.org/10.1016/j.rser.2015.05.024 (accessed on 20th February 2026).

Sivonen, K., Kononen, K., Carmichael, W. W., Dahlem, A. M., Rinehart, K. L., Kiviranta, J., et al. (1989). Occurrence of the hepatotoxic cyanobacterium *Nodularia spumigena* in the Baltic Sea and structure of the toxin. *Applied and Environmental Microbiology, 55*(8), 1990–1995. https://doi.org/10.1128/aem.55.8.1990-1995.1989 (accessed on 20th February 2026).

Smayda, T. J. (1990). Novel and nuisance phytoplankton blooms in the sea: Evidence for a global epidemic. In *Toxic Marine Phytoplankton* (pp. 29–40).

Smerilli, A., Orefice, I., Corato, F., Olea, A. G., Ruban, A. V., & Brunet, C. (2017). Photoprotective and antioxidant responses to light spectrum and intensity variations in the coastal diatom *Skeletonema marinoi*. *Environmental Microbiology, 19*(2), 611–627. https://doi.org/10.1111/1462-2920.13545 (accessed on 20th February 2026).

Smith, F. A., & Raven, J. A. (1979). Intracellular pH and its regulation. *Annual Review of Plant Physiology, 30*, 289–311.

Smith, R. C., Prézelin, B. B., Baker, K. S., Bidigare, R. R., Boucher, N. P., Coley, T., et al. (1992). Ozone depletion: Ultraviolet radiation and phytoplankton biology in Antarctic waters. *Science, 255*(5047), 952–959. https://doi.org/10.1126/science.1546292 (accessed on 20th February 2026).

Solovchenko, A., & Khozin-Goldberg, I. (2013). High-CO_2 tolerance in microalgae: Possible mechanisms and implications for biotechnology and bioremediation. *Biotechnology Letters, 35*(11), 1745–1752. https://doi.org/10.1007/s10529-013-1274-7 (accessed on 20th February 2026).

Solovchenko, A., Gorelova, O., Selyakh, I., Pogosyan, S., Baulina, O., Semenova, L., et al. (2015). A novel CO_2-tolerant symbiotic *Desmodesmus* (Chlorophyceae, Desmodesmaceae): Acclimation to and performance at a high carbon dioxide level. *Algal Research, 11*, 399–410. https://doi.org/10.1016/j.algal.2015.04.011 (accessed on 20th February 2026).

Sommer, U., & Lengfellner, K. (2008). Climate change and the timing, magnitude, and composition of the phytoplankton spring bloom. *Global Change Biology, 14*(6), 1199–1208. https://doi.org/10.1111/j.1365-2486.2008.01571.x (accessed on 20th February 2026).

Sommer, U., Peter, K. H., Genitsaris, S., & Moustaka-Gouni, M. (2017). Do marine phytoplankton follow Bergmann's rule sensu lato? *Biological Reviews, 92*(2), 1011–1026. https://doi.org/10.1111/brv.12266 (accessed on 20th February 2026).

Song, C., Dodds, W. K., Rüegg, J., Argerich, A., Baker, C. L., Bowden, W. B., et al. (2018). Continental-scale decrease in net primary productivity in streams due to climate warming. *Nature Geoscience, 11*(6), 415–420. https://doi.org/10.1038/s41561-018-0125-5 (accessed on 20th February 2026).

Song, H., Lavoie, M., Fan, X., Tan, H., Liu, G., Xu, P., et al. (2017). Allelopathic interactions of linoleic acid and nitric oxide increase the competitive ability of *Microcystis aeruginosa*. *The ISME Journal, 11*(8), 1865–1876. https://doi.org/10.1038/ismej.2017.45 (accessed on 20th February 2026).

Sonmez, C., Elcin, E., Akin, D., Oktem, H. A., & Yucel, M. (2016). Evaluation of novel thermo-resistant *Micractinium* and *Scenedesmus* sp. for efficient biomass and lipid production under different temperature and nutrient regimes. *Bioresource Technology, 211*, 422–428. https://doi.org/10.1016/j.biortech.2016.03.125 (accessed on 20th February 2026).

Spence, K. J., Digman, C., Balmforth, B., Houldsworth, J., Saul, A., & Meadowcroft, J. (2016). Gross solids from combined sewers in dry weather and storms, elucidating production, storage and social factors. *Urban Water Journal, 13*(8), 773–789. https://doi.org/10.1080/1573062X.2015.1025081 (accessed on 20th February 2026).

Srirangan, S., Sauer, M. L., Howard, B., Dvora, M., Dums, J., Backman, P., et al. (2015). Interaction of temperature and photoperiod increases growth and oil content in the marine microalga *Dunaliella viridis*. *PLoS ONE, 10*(5), e0127562. https://doi.org/10.1371/journal.pone.0127562 (accessed on 20th February 2026).

Stock, C. A., Dunne, J. P., & John, J. G. (2014). Drivers of trophic amplification of ocean productivity trends in a changing climate. *Bio Geosciences, 11*(6), 7125–7135. https://doi.org/10.5194/bg-11-7125-2014 (accessed on 20th February 2026).

Stock, C. A., John, J. G., Rykaczewski, R. R., & Watson, R. A. (2017). Reconciling fisheries catch and ocean productivity. *Proceedings of the National Academy of Sciences of the United States of America, 114*(8), E1441–E1449. https://doi.org/10.1073/pnas.1610238114 (accessed on 20th February 2026).

Straub, C., Quillardet, P., Vergalli, J., Tandeau de Marsac, N., & Humbert, J. F. (2011). A day in the life of *Microcystis aeruginosa* strain PCC 7806 as revealed by a transcriptomic analysis. *PLoS ONE, 6*(1), e16208. https://doi.org/10.1371/journal.pone.0016208 (accessed on 20th February 2026).

Stucken, K., John, U., Cembella, A., Murillo, A. A., & Soto-Liebe, K. (2010). The smallest known genomes of multicellular and toxic cyanobacteria: Comparison, minimal gene sets for linked traits and the evolutionary implications. *PLoS ONE, 5*(2), e9235. https://doi.org/10.1371/journal.pone.0009235 (accessed on 20th February 2026).

Stüken, A., Orr, R. J. S., Kellmann, R., Murray, S. A., Neilan, B. A., & Jakobsen, K. S. (2011). Discovery of nuclear-encoded genes for the neurotoxin saxitoxin in dinoflagellates. *PLoS ONE, 6*(5), e20096. https://doi.org/10.1371/journal.pone.0020096 (accessed on 20th February 2026).

Sun, J., Hutchins, D. A., Feng, Y., Seubert, E. L., Caron, D. A., & Fu, F. X. (2011). Effects of changing pCO_2 and phosphate availability on domoic acid production and physiology of the marine harmful bloom diatom *Pseudo-nitzschia multiseries*. *Limnology and Oceanography, 56*(3), 829–840. https://doi.org/10.4319/lo.2011.56.3.0829 (accessed on 20th February 2026).

Sun, M. M., Sun, J., Qiu, J. W., Jing, H., & Liu, H. (2012). Characterization of the proteomic profiles of the brown tide alga *Aureoumbra lagunensis* under phosphate- and nitrogen-limiting conditions and of its phosphate limitation-specific protein with alkaline phosphatase activity. *Applied and Environmental Microbiology, 78*(6), 2025–2033. https://doi.org/10.1128/AEM.05755-11 (accessed on 20th February 2026).

Suresh, K. K., Dahms, H. U., Won, E. J., Lee, J. S., & Shin, K. H. (2015). Microalgae—A promising tool for heavy metal remediation. *Ecotoxicology and Environmental Safety, 113*, 329–352.

Swarnalatha, G. V., Hegde, N. S., Chauhan, V. S., & Sarada, R. (2015). The effect of carbon dioxide rich environment on carbonic anhydrase activity, growth and metabolite production in indigenous freshwater microalgae. *Algal Research, 9*, 151–159. https://doi.org/10.1016/j.algal.2015.02.014 (accessed on 20th February 2026).

Tagliaferro, M., Gonçalves, A. M. M., Bergman, M., Sobral, O., & Graça, M. A. S. (2018). Assessment of metal exposure (uranium and copper) by the response of a set of integrated biomarkers in a stream shredder. *Ecological Indicators, 95*, 991–1000. https://doi.org/10.1016/j.ecolind.2017.10.065 (accessed on 20th February 2026).

Takahashi, S., Nakajima, N., Saji, H., & Kondo, N. (2002). Diurnal change of cucumber CPD photolyase gene (*CsPHR*) expression and its physiological role in growth under UV-B irradiation. *Plant and Cell Physiology, 43*(3), 342–349. https://doi.org/10.1093/pcp/pcf038 (accessed on 20th February 2026).

Tan, Y. H., Lim, P. E., Beardall, J., Poong, S. W., & Phang, S. M. (2019). A metabolomic approach to investigate effects of ocean acidification on a polar microalga *Chlorella* sp. *Aquatic Toxicology, 217*, 105349. https://doi.org/10.1016/j.aquatox.2019.105349 (accessed on 20th February 2026).

Tatters, A. O., Flewelling, L. J., Fu, F., Granholm, A. A., & Hutchins, D. A. (2013). High CO_2 promotes the production of paralytic shellfish poisoning toxins by *Alexandrium catenella* from southern California waters. *Harmful Algae, 30*, 37–43. https://doi.org/10.1016/j.hal.2013.08.007 (accessed on 20th February 2026).

Teegarden, G. J., & Cembella, A. D. (1996). Grazing of toxic dinoflagellates, *Alexandrium* spp., by adult copepods of coastal Maine: Implications for the fate of paralytic shellfish toxins in marine food webs. *Journal of Experimental Marine Biology and Ecology, 196*, 145–176. https://doi.org/10.1016/0022-0981(95)00128-X (accessed on 20th February 2026).

Teng, H., Washington, W. M., Meehl, G. A., Buja, L. E., & Strand, G. W. (2006). Twenty-first century Arctic climate change in the CCSM3 IPCC scenario simulations. *Climate Dynamics, 26*(6), 601–616. https://doi.org/10.1007/s00382-005-0099-z (accessed on 20th February 2026).

Teng, Y. C., Primeau, F. W., Moore, J. K., Lomas, M. W., & Martiny, A. C. (2014). Global-scale variations of the ratios of carbon to phosphorus in exported marine organic matter. *Nature Geoscience, 7*(12), 895–898. https://doi.org/10.1038/ngeo2303 (accessed on 20th February 2026).

Tester, P. A., Litaker, R. W., & Berdalet, E. (2020). Climate change and harmful benthic microalgae. *Harmful Algae, 91*, 101655. https://doi.org/10.1016/j.hal.2019.101655 (accessed on 20th February 2026).

Thomas, M. K., Kremer, C. T., Klausmeier, C. A., & Litchman, E. (2012). A global pattern of thermal adaptation in marine phytoplankton. *Science, 338*(6107), 1085–1088. https://doi.org/10.1126/science.1224836 (accessed on 20th February 2026).

Thornhill, D. J., LaJeunesse, T. C., Kemp, D. W., Fitt, W. K., & Schmidt, G. W. (2006). Multi-year, seasonal genotypic surveys of coral–algal symbioses reveal prevalent stability or post-bleaching reversion. *Marine Biology, 148*(4), 711–722. https://doi.org/10.1007/s00227-005-0114-2 (accessed on 20th February 2026).

Thorsen, T. A., & Dale, B. (1997). Dinoflagellate cysts as indicators of pollution and past climate in a Norwegian fjord. *The Holocene, 7*(4), 433–446. https://doi.org/10.1177/095968369700700406 (accessed on 20th February 2026).

Tichadou, L., Glaizal, M., Armengaud, A., Grossel, H., Lemée, R., Kantin, R., et al. (2010). Health impact of unicellular algae of the genus *Ostreopsis* blooms in the Mediterranean Sea: Experience of the French Mediterranean coast surveillance network from 2006 to 2009. *Clinical Toxicology, 48*(8), 839–844. https://doi.org/10.3109/15563650.2010.513687 (accessed on 20th February 2026).

Tilzer, M. M., Elbrächter, M., Gieskes, W. W., & Beese, B. (1986). Light–temperature interactions in the control of photosynthesis in Antarctic phytoplankton. *Polar Biology, 5*(2), 105–111. https://doi.org/10.1007/BF00443382 (accessed on 20th February 2026).

Tittensor, D. P., Eddy, T. D., Lotze, H. K., Galbraith, E. D., Cheung, W., Barange, M., et al. (2018). A protocol for the intercomparison of marine fishery and ecosystem models: Fish-MIP v1.0. *Geoscientific Model Development, 11*(4), 1421–1442. https://doi.org/10.5194/gmd-11-1421-2018 (accessed on 20th February 2026).

Tortell, P. D. (2000). Evolutionary and ecological perspectives on carbon acquisition in phytoplankton. *Limnology and Oceanography, 45*(3), 744–750. https://doi.org/10.4319/lo.2000.45.3.0744 (accessed on 20th February 2026).

Trainer, V. L., Eberhart, B. T. L., Wekell, J. C., Adams, N. G., Hanson, L., Cox, F., et al. (2003). Paralytic shellfish toxins in Puget Sound, Washington State. *Journal of Shellfish Research, 22*(1), 213–223.

Tripp, H. J., Bench, S. R., Turk, K. A., Foster, R. A., Desany, B. A., Niazi, F., et al. (2010). Metabolic streamlining in an open-ocean nitrogen-fixing cyanobacterium. *Nature, 464*(7285), 90–94. https://doi.org/10.1038/nature08786 (accessed on 20th February 2026).

Tubaro, A., Durando, P., Del Favero, G., Ansaldi, F., Icardi, G., Deeds, J. R., et al. (2011). Case definitions for human poisonings postulated to palytoxin exposure. *Toxicon, 57*(3), 478–495. https://doi.org/10.1016/j.toxicon.2011.01.005 (accessed on 20th February 2026).

United Nations (UN). (2015). *Water for a Sustainable World Report.*

Van Baalen, C., & O'Donnell, R. (1983). Isolation and growth of psychrophilic diatoms from the ice-edge in the Bering Sea. *Microbiology, 129*(4), 1019–1023. https://doi.org/10.1099/00221287-129-4-1019 (accessed on 20th February 2026).

Van de Poll, W. H., Eggert, A., Buma, A. G. J., & Breeman, A. M. (2001). Effects of UV-B-induced DNA damage and photoinhibition on growth of temperate marine red macrophytes: Habitat-related differences in UV-B tolerance. *Journal of Phycology, 37*(1), 30–38. https://doi.org/10.1046/j.1529-8817.2001.037001030.x (accessed on 20th February 2026).

Van de Poll, W. H., Van Leeuwe, M. A., Roggeveld, J., & Buma, A. G. J. (2005). Nutrient limitation and high irradiance acclimation reduce PAR and UV-induced viability loss in the Antarctic diatom *Chaetoceros brevis* (Bacillariophyceae). *Journal of Phycology, 41*(4), 840–850. https://doi.org/10.1111/j.1529-8817.2005.00105.x (accessed on 20th February 2026).

Van de Waal, D. B., Eberlein, T., John, U., Wohlrab, S., & Rost, B. (2014). Impact of elevated pCO_2 on paralytic shellfish poisoning toxin content and composition in *Alexandrium tamarense*. *Toxicon, 78*, 58–67. https://doi.org/10.1016/j.toxicon.2013.11.011 (accessed on 20th February 2026).

Van den Bergh, B., Toon, S., Maarten, F., & Jan, M. (2018). Experimental design, population dynamics, and diversity. *Applied and Environmental Microbiology, 82*(3), 1–54.

Van Dolah, F. M., Lidie, K. B., Morey, J. S., Brunelle, S. A., Ryan, J. C., Monroe, E. A., et al. (2007). Microarray analysis of diurnal- and circadian-regulated genes in the Florida red-tide dinoflagellate *Karenia brevis* (Dinophyceae). *Journal of Phycology, 43*(4), 741–752. https://doi.org/10.1111/j.1529-8817.2007.00354.x (accessed on 20th February 2026).

Vardi, A., Formiggini, F., Casotti, R., & De Martino, A. (2006). A stress surveillance system based on calcium and nitric oxide in marine diatoms. *PLoS Biology, 4*(3), e60. https://doi.org/10.1371/journal.pbio.0040060 (accessed on 20th February 2026).

Varjani, S., Agarwal, A. K., Gnansounou, E., & Gurunathan, B. (2018). *Bioremediation: Applications for Environmental Protection and Management.* Springer.

Vila, M., Masó, M., Sampedro, N., et al. (2008). The genus *Ostreopsis* in recreational waters of the Catalan coast and Balearic Islands (NW Mediterranean Sea): Is this the origin of human respiratory difficulties? In *Proceedings of the 12th International Conference on Harmful Algae* (pp. 334–336).

Voß, B., Bolhuis, H., Fewer, D. P., Kopf, M., Möke, F., Haas, F., et al. (2013). Insights into the physiology and ecology of the brackish-water-adapted cyanobacterium *Nodularia spumigena* CCY9414 based on a genome-transcriptome analysis. *PLoS ONE, 8*(3), e60224. https://doi.org/10.1371/journal.pone.0060224 (accessed on 20th February 2026).

Wang, D. Z. (2008). Neurotoxins from marine dinoflagellates: A brief review. *Marine Drugs, 6*(2), 349–371. https://doi.org/10.3390/md20080016 (accessed on 20th February 2026).

Wang, D. Z., Lin, L., Wang, M. H., Li, C., & Hong, H. S. (2012). Proteomic analysis of a toxic dinoflagellate *Alexandrium catenella* under different growth phases and conditions. *Chinese Science Bulletin, 57*(25), 3328–3341. https://doi.org/10.1007/s11434-012-5160-9 (accessed on 20th February 2026).

Wang, D. Z., Zhang, Y. J., Zhang, S. F., Lin, L., & Hong, H. S. (2013). Quantitative proteomic analysis of the cell cycle of the dinoflagellate *Prorocentrum donghaiense* (Dinophyceae). *PLoS ONE, 8*(5), e63659. https://doi.org/10.1371/journal.pone.0063659 (accessed on 20th February 2026).

Wang, J. H., Zhang, T. Y., Dao, G. H., Xu, X. Q., Wang, X. X., & Hu, H. Y. (2017). Microalgae-based advanced municipal wastewater treatment for reuse in water bodies. *Applied Microbiology and Biotechnology, 101*(7), 2659–2675.

Wang, Y., Duanmu, D., & Spalding, M. H. (2011). Carbon dioxide concentrating mechanism in *Chlamydomonas reinhardtii*: Inorganic carbon transport and CO_2 recapture. *Photosynthesis Research, 109*(1–3), 115–122. https://doi.org/10.1007/s11120-011-9643-3 (accessed on 20th February 2026).

Waterworth, W. M. (2002). Characterization of *Arabidopsis* photolyase enzymes and analysis of their role in protection from ultraviolet-B radiation. *Journal of Experimental Botany, 53*(371), 1005–1015. https://doi.org/10.1093/jexbot/53.371.1005 (accessed on 20th February 2026).

Wells, M. L., Trainer, V. L., Smayda, T. J., Karlson, B. S. O., Trick, C. G., Kudela, R. M., et al. (2015). Harmful algal blooms and climate change: Learning from the past and present to forecast the future. *Harmful Algae, 49*, 68–93. https://doi.org/10.1016/j.hal.2015.07.009 (accessed on 20th February 2026).

Wetzel, R. G. (2001). Planktonic communities: Algae and cyanobacteria. In *Limnology* (pp. 331–393). https://doi.org/10.1016/B978-0-08-057439-4.50019-8 (accessed on 20th February 2026).

Weydmann, A., Søreide, J. E., Kwaśniewski, S., & Widdicombe, S. (2012). Influence of CO_2-induced acidification on the reproduction of a key Arctic copepod *Calanus*

glacialis. *Journal of Experimental Marine Biology and Ecology, 428*, 39–42. https://doi.org/10.1016/j.jembe.2012.06.002 (accessed on 20th February 2026).
White, A. W. (1981). Marine zooplankton can accumulate and retain dinoflagellate toxins and cause fish kills. *Limnology and Oceanography, 26*(1), 103–109. https://doi.org/10.4319/lo.1981.26.1.0103 (accessed on 20th February 2026).
Wigley, T. M. L., & Raper, S. C. B. (2001). Interpretation of high projections for global-mean warming. *Science, 293*(5529), 451–454. https://doi.org/10.1126/science.1061604 (accessed on 20th February 2026).
Winck, F. V., Melo, D. O. P., Riaño-Pachón, D. M., Martins, M. C. M., Caldana, C., & Barrios, A. F. G. (2016). Analysis of sensitive CO_2 pathways and genes related to carbon uptake and accumulation in *Chlamydomonas reinhardtii* through genomic scale modeling and experimental validation. *Frontiers in Plant Science, 7*, 43. https://doi.org/10.3389/fpls.2016.00043 (accessed on 20th February 2026).
Wohlrab, S., Iversen, M. H., & John, U. (2010). A molecular and co-evolutionary context for grazer-induced toxin production in *Alexandrium tamarense*. *PLoS ONE, 5*(11), e15039. https://doi.org/10.1371/journal.pone.0015039 (accessed on 20th February 2026).
Wollmann, F., Dietze, S., Ackermann, J. U., Bley, T., Walther, T., Steingroewer, J., et al. (2019). Microalgae wastewater treatment: Biological and technological approaches. *Engineering in Life Sciences, 19*(12), 860–871. https://doi.org/10.1002/elsc.201900071 (accessed on 20th February 2026).
Wright, J. L. C., Boyd, R. K., de Freitas, A. S. W., Falk, M., Foxall, R. A., Jamieson, W. D., et al. (1989). Identification of domoic acid, a neuroexcitatory amino acid, in toxic mussels from eastern Prince Edward Island. *Canadian Journal of Chemistry, 67*(3), 481–490. https://doi.org/10.1139/v89-075 (accessed on 20th February 2026).
Wu, P., Zakem, E. J., Dutkiewicz, S., & Zhang, Y. (2020). Biomagnification of methylmercury in a marine plankton ecosystem. *Environmental Science & Technology*. https://doi.org/10.1021/acs.est.9b06075 (accessed on 20th February 2026).
Wu, Y., Guan, K., Wang, Z., Xu, B., & Zhao, F. (2013). Isolation, identification and characterization of an electrogenic microalgae strain. *PLoS ONE, 8*(9), e73442. https://doi.org/10.1371/journal.pone.0073442 (accessed on 20th February 2026).
Wurch, L. L., Bertrand, E. M., Saito, M. A., van Mooy, B. A. S., & Dyhrman, S. T. (2011). Proteome changes driven by phosphorus deficiency and recovery in the brown tide-forming alga *Aureococcus anophagefferens*. *PLoS ONE, 6*(12), e28949. https://doi.org/10.1371/journal.pone.0028949 (accessed on 20th February 2026).
Xavier, J. C., Brandt, A., Ropert-Coudert, Y., Badhe, R., Gutt, J., Havermans, C., et al. (2016). Future challenges in Southern Ocean ecology research. *Frontiers in Marine Science, 3*, 94. https://doi.org/10.3389/fmars.2016.00094 (accessed on 20th February 2026).
Xia, B., Sui, Q., Sun, X., Han, Q., Chen, B., Zhu, L., et al. (2018). Ocean acidification increases the toxic effects of TiO_2 nanoparticles on the marine microalga *Chlorella vulgaris*. *Journal of Hazardous Materials, 346*, 1–37. https://doi.org/10.1016/j.jhazmat.2017.12.017 (accessed on 20th February 2026).
Xu, P., & Huang, L. (2017). Stereoselective bioaccumulation, transformation, and toxicity of triadimefon in *Scenedesmus obliquus*. *Chirality, 29*(2), 61–69. https://doi.org/10.1002/chir.22671 (accessed on 20th February 2026).

Yang, I., Beszteri, S., Tillmann, U., Cembella, A., & John, U. (2011). Growth- and nutrient-dependent gene expression in the toxigenic marine dinoflagellate *Alexandrium minutum*. *Harmful Algae, 12*, 55–69. https://doi.org/10.1016/j.hal.2011.08.012 (accessed on 20th February 2026).

Yue, D., Peng, Y., Yin, Q., & Xiao, L. (2015). Proteomic analysis of *Microcystis aeruginosa* in response to nitrogen and phosphorus starvation. *Journal of Applied Phycology, 27*(3), 1195–1204. https://doi.org/10.1007/s10811-014-0405-4 (accessed on 20th February 2026).

Zeppilli, D., Sarrazin, J., Leduc, D., Martinez Arbizu, P., Fontaneto, D., Fontanier, C., et al. (2015). Is the meiofauna a good indicator for climate change and anthropogenic impacts? *Marine Biodiversity, 45*(3), 505–535. https://doi.org/10.1007/s12526-015-0359-z (accessed on 20th February 2026).

Zhang, Q., & Rickaby, R. E. M. (2020). Interactions of thallium with marine phytoplankton. *Geochimica et Cosmochimica Acta, 276*, 1–13. https://doi.org/10.1016/j.gca.2020.02.024 (accessed on 20th February 2026).

Zhang, S. F., Zhang, Y., Xie, Z. X., Zhang, H., Lin, L., & Wang, D. Z. (2015). iTRAQ-based quantitative proteomic analysis of a toxigenic dinoflagellate *Alexandrium catenella* and its non-toxic mutant. *Proteomics, 15*(9), 4041–4050. https://doi.org/10.1002/pmic.201500156 (accessed on 20th February 2026).

Zingone, A., & Enevoldsen, H. O. (2000). The diversity of harmful algal blooms: A challenge for science and management. *Ocean & Coastal Management, 43*(8), 725–748. https://doi.org/10.1016/S0964-5691(00)00056-9 (accessed on 20th February 2026).

CHAPTER 8

Prospective Applications of Microalgae as a Model for Ecotoxicity Testing

PRABHAKAR MISHRA,* DEITY ANAND, and PADMASHREE SEKAR

Department of Biotechnology, School of Applied Sciences, REVA University, Bangalore, Karnataka, India

*Corresponding author

ABSTRACT

Microalgae are eukaryotic, unicellular, microscopic species that contain chloroplasts and undergo photosynthesis, contributing to ecological balance. Microalgae are commonly used in ecotoxicity testing due to their ease of cultivation and rapid cell division rates. These properties make microalgal species suitable for toxicity studies. Changes in the morphological or chemical properties of microalgae determine the toxic impact of pollutants in an ecosystem. Many microalgal species are adapted for use in different toxicity assessments. Microalgal species are primary organisms in aquatic ecosystems, which makes them highly interactive with various forms of pollutants, ranging from macro- to nanopollutants. Microalgae have been exploited as model microorganisms in numerous ecotoxicity studies due to their physiological mechanisms, which include membrane permeability, osmoregulation, transition to multicellularity, and tolerance. In addition to these valuable features, microalgae possess diversified application potential in industrial contexts. This chapter summarizes the potential of these algal systems from a multidimensional perspective.

Environmental Phycology: Climate Change Mitigation and Hazardous Pollutant Cleanup Using Algal Resources. Jeyabalan Sangeetha and Devarajan Thangadurai (Eds.)

DOI: 10.1201/9781779647085-8

8.1 INTRODUCTION

Microalgae are eukaryotic, unicellular, microscopic species that exist individually or in groups. These microorganisms have diversified application potential (Figure 8.1). They are similar to green plants; that is, they contain chloroplasts and undergo photosynthesis by using CO_2, water, and sunlight to produce their own food, with the exception of *Euglena*, which does not contain chloroplasts and ingests other organisms for food. Microalgae are found in both marine and freshwater environments. They are extremely diverse and occur in nearly every habitat, serving as the base of aquatic food chains (Cid et al., 1995; Kumar and Buddolla, 2019). The evolution of algae began approximately 1.5 billion years ago, and cyanobacteria are considered the originators of photosynthesis and the ancestors of all eukaryotic algae. Algae are highly diverse and are usually classified based on their photosynthetic pigments, such as green, yellow, red, and brown. Some microalgae include cells derived from a primary endosymbiotic event. During this event, a highly reduced cyanobacterium-derived genome resulted in a photosynthetic plastid surrounded by two membranes; this process is known as endosymbiosis. The dominant marine eukaryotic algae possess plastids originating from the engulfment of red algae derived from secondary endosymbiotic events. These include diatoms, dinoflagellates, haptophytes, and cryptomonads. When the original plastid from secondary red algal endosymbiosis is discarded in favor of a replacement symbiont, this is termed tertiary endosymbiosis. For example, the peridinin plastid derived from a haptophyte was replaced in the dinoflagellate *Karenia brevis* (Figure 8.2).

The reproductive phase in microalgae involves both asexual and sexual methods. Asexual reproduction occurs through the production of spores, which may be flagellated (motile). Sexual reproduction takes place through the fusion of gametes, which can be either flagellated (motile) or nonflagellated (nonmotile).

The biggest advantage is that the growth rate of microalgae is extremely fast, with high lipid content, using CO_2 as a substrate. They are primary producers and are found to be beneficial to human beings. They are produced on a limited scale for high-value products such as polyunsaturated fatty acids, food supplements, and natural pigments. The energy required to produce biomass from microalgae is very high. Globally, many different organizations prefer the production of biodiesel

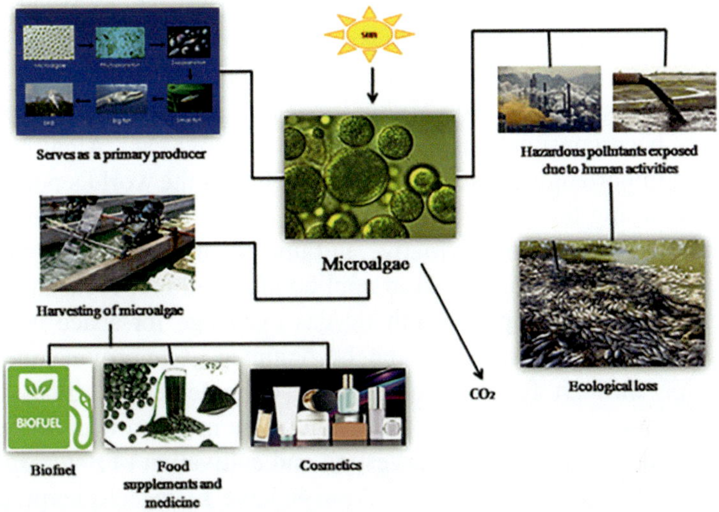

FIGURE 8.1 Applications of microalgae as an ecotoxicity testing model.

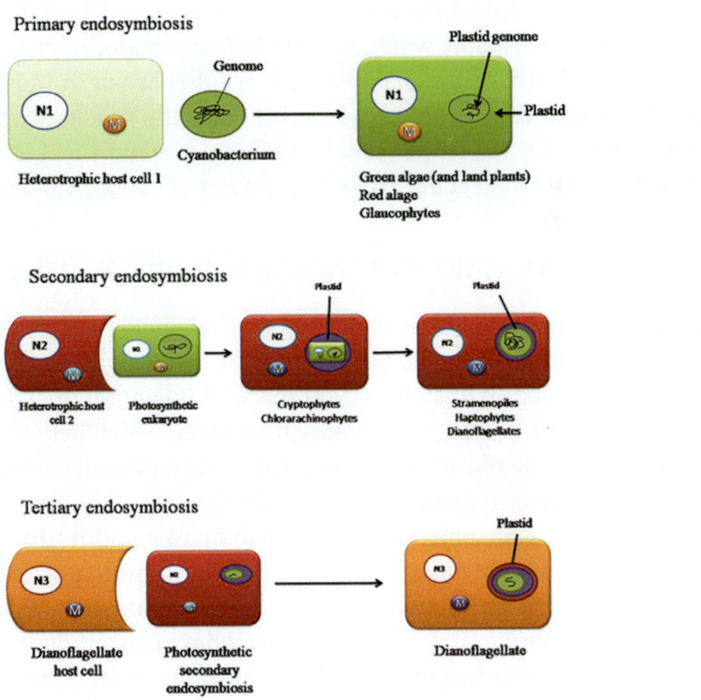

FIGURE 8.2 Illustration of primary, secondary, and tertiary endosymbiosis in microalgae.

through a process known as transesterification. Applications are based on environmental remediation and drug development (Ray et al., 2019).

Indigenous populations have used microalgae for centuries. There are edible algae, such as *Nostoc*, *Aphanizomenon*, and *Arthrospira* species, that have been consumed as food for thousands of years. Algal biomass served an important purpose in the early 1950s. As the world's population increased during this period, predictions of an insufficient protein supply led to the exploration of alternative and unconventional protein sources. Studies subsequently continued in applied algal culture, focusing on the use of microalgae for photosynthetic gas exchange for space travelers. Environmental technologies in the USA aimed at improving methane production through fermentation were also implemented. In the early 1960s, large-scale commercial production began in Japan with the culture of *Chlorella*, followed by the harvesting and cultivation of *Arthrospira* in the early 1970s. During the 1980s in Asia, 46 large-scale factories produced more than 1000 kg of microalgal biomass (*Chlorella*) per month. In 1986, Betatene (Whyalla, Australia) and Western Biotechnology (Hutt Lagoon, Australia) established the commercial production of *Dunaliella salina* as a source of β-carotene, which became the third major microalgal industry. Commercial production of cyanobacteria also began in India during this period. The microalgal biomass market currently produces approximately 5000 tons of dry matter per year and generates about US$1.25 × 10^9 annually (Ramakrishnan et al., 2010; Spolaore et al., 2006; Talukder, 2014).

8.2 MICROALGAE IN THE AQUATIC ENVIRONMENT

Microalgae are ubiquitous and are found in aquatic environments. There are about 200,000 to 800,000 species of microalgae in many different genera, with approximately 30,000 species known. Microalgae are important consumers of CO_2 and primary producers, which help to stabilize the food chain in aquatic environments. They also represent converters of solar energy to biomass. There are few microalgae found in marine water, few in freshwater, and some that live in both marine and freshwater habitats. The photosynthetic microbial representatives from prokaryotic organisms, such as cyanobacteria, and eukaryotic organisms, such as diatoms, concerning marine and freshwater, are known to produce a myriad of toxic or bioactive metabolites (Franklin et al., 2001; Vandamme et al., 2011). There are photosynthetic lineages derived from a single primary endosymbiotic event: the green algae, also known as Chlorophyceae. There are approximately 8000 species of green algae. Their cell walls consist of cellulose. They are green in color due to the dominance of the pigments chlorophyll *a* and *b* (Eullaffroy and Vernet, 2003). Most green algae contain pyrenoids that act as storage bodies, storing starch, and some store food in the form of oil droplets. They are usually found in freshwater, brackish water, and marine environments. Rhodophyceae, also known as red algae, have *R*-phycoerythrin present in their cells, which is responsible for the predominance of the red pigment; there are approximately 7000 species of red algae. They also contain chlorophyll *a* and *d*. They are found in high concentrations in warmer regions and marine waters, with few species in freshwater and brackish water. Their cell walls consist of cellulose, pectin, and polysulfate esters. Phaeophyceae, also known as brown algae, are commonly found in marine habitats, with approximately 1500–2000 known species. They contain chlorophyll *a*, *c*, and fucoxanthin. The pigments xanthophylls give them their brown and olive-green colors, depending on the amount of xanthophylls present. Their cell walls consist of cellulose and algin.

8.2.1 MARINE AQUATIC ENVIRONMENT

As recorded, marine algae are responsible for 40–50% of the photosynthesis that occurs on Earth every year. The photosynthesis of the marine aquatic region is dominated by microalgae, which, together with cyanobacteria, are called phytoplankton. The marine algae genomics was first published in the 551-kilobase remnant nucleomorph genomics of the cryptomonad *Guillaridia theta.* In many marine invertebrates, including mollusks and sponges, microalgae are found on them. These photosynthetic symbionts provide efficient nutrients to their hosts. For example, in the case of corals, the microalgae, also called zooxanthellae, provide over 50% of the carbon and energy requirements of their hosts (Berry et al., 2007).

8.2.2 FRESHWATER AQUATIC ENVIRONMENT

Freshwater microalgae play an important role in the aquatic environment. As we know, freshwater algae and microalgae produce molecular oxygen, which is essential, but their metabolic by-product, known as reactive oxygen species (ROS), is highly toxic and leads to oxidative stress. Using the antioxidant response mechanism, freshwater microalgae can prevent the effects of ROS. The mechanisms performed by freshwater microalgae vary from species to species based on the following factors: cell size, cell shape, cell density, growth stage, and a few abiotic factors, such as temperature, light, and nutrients. Freshwater microalgae also contribute to the biodegradation or biotransformation of BPA in aquatic environments. The surface ratio of freshwater microalgae species is slightly higher than that of marine water microalgae species (Berry et al., 2007; Dazy et al., 2009; Kellam et al., 1988).

8.2.3 DIATOMS

Diatoms are eukaryotic algae that play a very important role in the global cycling of carbon and oxygen and form the basis of aquatic food webs. Their cell walls are embedded with silica and consist of two thin layers that overlap and fit together like a soapbox. Hence, diatoms deposit large amounts of silica-based cell wall material in their habitats; this accumulation over billions of years is known as "diatomaceous earth." Diatoms occur in both marine and freshwater habitats with respect to salinity; most diatom genera are restricted to either marine or freshwater environments. Diatoms are also powerful indicators of environmental change. The deposits in which diatoms are found can be used for polishing and for the filtration of syrups and oils. Therefore, diatoms are also known as the "chief producers" of the ocean.

8.3 APPLICATIONS OF MICROALGAE

8.3.1 MICROALGAE IN HUMAN NUTRITION

Microalgae in different forms, such as capsules, tablets, syrups, and so forth, are marketed as a source of nutrition. They contain proteins, carbohydrates, vitamins A, C, B1, B2, and B6, iodine, potassium, magnesium, calcium, fiber, and enzymes. *Arthrospira*, *Chlorella*, *Dunaliella salina*, and *Aphanizomenon flos-aquae* are the main genera used in commercial applications (Spolaore et al., 2006). *Arthrospira* is widely produced in India and China because it has a high protein content and serves as excellent nutrition for humans. Additionally, microalgae have health-promoting effects, such as suppression of hypertension, protection against renal failure, and alleviation of hyperlipidemia. Earthrise Farma, located in Calipatria, CA, USA, has the largest production plant in the world, and *Arthrospira* is distributed in over 20 countries. *Chlorella* contains an important component, β-1,3-glucan, which is a free-radical scavenger, an active immunostimulator, and reduces blood lipid levels. *Dunaliella salina* is important for its β-carotene content, which accounts for about 14% of its dry weight (Priyadarshani and Rath, 2012). For human consumption, strains of *Dunaliella salina* are produced as powder and used as an ingredient in dietary supplements by Cognis Nutrition and Health, the world's largest producer of this strain. *Aphanizomenon flos-aquae* promotes overall health and plays a vital role in commercial strain applications (Sousa et al., 2008; Spolaore et al., 2006) (Table 8.1 and Figure 8.3).

TABLE 8.1 Major Microalgae Commercialized for Human Nutrition

Microalgae	Major Producers	Products	References
Aphanizomenon flos-aquae	Blue Green Foods (USA), Vision (USA)	Capsules, crystals, and powder	Spolaore et al. (2006)
Chlorella	Taiwan Chlorella Manufacturing Co. (Taiwan), Klotze (Germany)	Tablets, powders, nectar, and noodles	Priyadarshani et al. (2012)
Dunaliella salina	Cognis Nutrition and Health (Australia)	Powder and carotene	Priyadarshani et al. (2012)
Spirulina (*Arthrospira*)	Cyanotech Corp. (Hawaii, USA), Myanmar Spirulina Factory (Myanmar)	Tablets, powders, beverages, extracts, chips, pasta, and liquid extract	Hallmann et al. (2007)

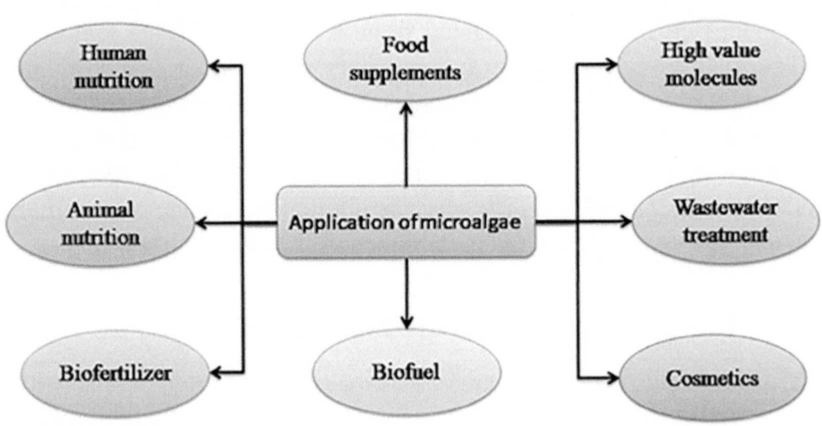

FIGURE 8.3 Applications of microalgae.

8.3.2 ANIMAL NUTRITION

A wide variety of animals are fed using microalgae, ranging from aquaculture to pets or farm animals as supplements. In fact, about 30% of the current world algal production is utilized for animal feed applications. For aquaculture species, the main application of microalgae is associated with nutrition, being fresh, and for inducing several other biological activities. The main nutritional value is determined by the protein complex. The highly unsaturated fatty acids such as EPA (eicosapentaenoic acid), AA (arachidonic acid), and DHA (docosahexaenoic acid) are important at their absolute levels. Vitamins from microalgae are also taken into account as being equally important. These factors provide balanced nutrition and improvement in animal growth. *Arthrospira* is again largely used in the domains of animal feed applications such as for aquarium fish, cows, dogs, horses, breeding bulls, and ornamental fish. Microalgae in animals provide a large profile of minerals, natural vitamins, improved immune systems, essential fatty acids, and fertility. A microalgae supplement is not fed in any concentration and is given in limited concentrations according to each variety. If not maintained, the animal may face adverse effects.

8.3.3 MICROALGAE IN COSMETICS

Cosmetic products related to skin care, facial care, refreshing, regenerating, and other applications contain extracts of microalgae. Microalgae also play a vital role in sun protection and hair care products. For example, a product rich in protein extracted from *Arthrospira* repairs skin aging before visible signs develop and prevents stria formation (Protulines, Exsymol SAM, Monaco), and a product containing an extract of *Chlorella vulgaris* supports collagen synthesis in the skin; therefore, it aids in tissue regeneration and helps reduce wrinkles (Dermochlorella, CODIF, St. Malo, France). *Dunaliella salina* has the ability to stimulate cell proliferation and influences the energy metabolism of the skin (Pepha®-Ctive).

8.3.4 MICROALGAE AND HIGH-VALUE MOLECULES

Microalgae represent untapped reservoirs of novel compounds, which are a very interesting natural source of new compounds with biological activity that could serve as functional ingredients. Therefore, it is known that microalgae also live in a complex habitat with extreme conditions, such as climate change in temperature, nutrients, salinity, and UV radiation. These factors should be noted when microalgae adapt to new environmental conditions to survive, which indeed produces great metabolites that are usually absent in other organisms.

8.3.5 BIOFUEL

Microalgae are an excellent source for biofuel production due to their high oil content and biomass production. The storage of lipids (triglycerides) for some microalgae can range up to 70% dry weight under nitrogen starvation. *Chlorella protothecoides* is known to be a potential feedstock for biodiesel production, with a crude lipid content of 55.2% of dry weight (grown under heterotrophic conditions with glucose). Hence, microalgae could provide feedstock for renewable liquid fuels, such as bioethanol and biodiesel (Enamala et al., 2018).

8.3.6 BIOFERTILIZER

Microalgae, such as cyanobacteria, play an important role in agriculture as soil conditioners and biofertilizers; that is, cyanobacteria act as a natural biofertilizer that builds up soil fertility and consequently increases rice growth. A cyanobacterium can fix nitrogen and help in the good cultivation of rice. Apart from the increase in yield and savings on fertilizer nitrogen, the soil physicochemical properties are also improved, such as pH, electrical conductivity, and in terms of grain quality, there is an improvement in protein content. Many varieties of cyanobacteria, or blue-green algae, have been recognized for the benefit of crop plants by providing several growth-promoting substances (Priyadarshani and Rath, 2012).

8.3.7 APPLICATIONS IN ENVIRONMENTAL BIOTECHNOLOGY

The cultivation of many microalgae is known to be the most profitable business in the field of the biotechnology industry, as it is an ecologically pure, energy- and resource-saving process that is waste-free. Under the category of biotechnology, applications are based on bioassay, biomonitoring, and bioremediation of environmental toxicity. The most efficient system to treat wastewater is the high-rate algae pond (HRAP). HRAP plays a very important role in the tertiary treatment of wastewater before the discharge of the water after secondary treatment. *Chlorella vulgaris* grown in HRAP has been shown to be the most significant for the final polishing of textile wastewater, especially the removal of color before discharge. Apart from the reduction of pollutants such as biological oxygen demand (BOD), chemical oxygen demand (COD), phosphorus, and nitrogen, HRAP generates algal biomass that has high applications in animal feed and as feedstock for biodiesel (Chu, 2012).

Bioassays evaluate the toxicity of pollutants using microalgae, including pesticides, heavy metals, and pharmaceuticals. In the Selangor River, due to intense urban activities, pesticides such as organochlorines and organophosphates have been detected. In aquatic systems, marine organisms such as fish and mussels have been found to contain heavy metals. The excessive use of pharmaceuticals, such as antibiotics in aquaculture and animal husbandry, has been observed in recent years. The most common microalgae used for bioassays of toxicants include *Dunaliella tertiolecta*, *Pseudokirchneriella subcapitata*, *Isochrysis galbana*, and *Chlorella* species. Various factors are used to assess the toxicity of pollutants to microalgae, including oxidative stress markers, photosynthesis, effects on growth, movement behavior, pigmentation, and biochemical composition (Bossuyt et al., 2004; Sigamani et al., 2016).

8.4 NATURAL MICROALGAL PREDATORY RELATIONSHIPS

Just like other living organisms, microalgae do have predators. Predators are necessary for microalgae as they balance the ecosystem by eliminating microalgal toxic blooms, but sometimes, when the ratio of predators increases beyond normal, it could lead to a decrease in the population of microalgae, which in turn results in reduced production as a primary producer and loss of biomass (Grami et al., 2011). The main effect of pathogens is the decrease in the growth rate of microalgae, and sometimes it can lead to death. Some of the predators include grazers, fungi, bacteria, protozoans, dinoflagellates, nano-flagellates, zooplankton, and viruses (Figure 8.4). Protozoans like Amoeba attack microalgal cell walls using a spike-like structure to suck out or extract cellular content. Pathogens like heterotrophic dinoflagellates and nano-flagellates feed on other microalgae, such as diatoms (Gachon et al., 2010; Simkovsky et al., 2012). Another group of pathogens known as zooplankton, which are multicellular organisms like crustaceans (copepods), fall under the predators category (Carney and Lane, 2014).

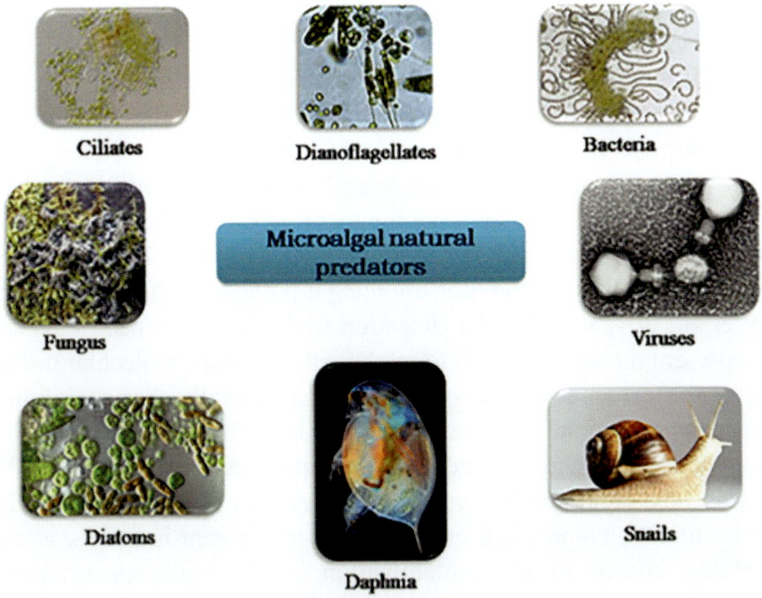

FIGURE 8.4 Natural predators of microalgae.

Most microalgal pathogens are highly prevalent; hence, over successive generations, some microalgae have developed defense mechanisms against pathogens and predators. Morphologically, cell size plays an important role: when microalgae are single-celled, they are more easily attacked or ingested by predators, whereas when they form colonies or clumps, predators may find them more difficult to ingest. Similar to plants, microalgae also possess sheaths and spines; for example, *Phormidium* develops sheaths that make it difficult for ciliates to ingest, and *Scenedesmus* sp. contains spines that help defend against predators. Another defense mechanism involves the development of thicker cell walls, which resist digestion by predators that feed on them. Chemical defenses include the production of toxins (e.g., DMS, microcystins) and repellents. Defenses may also be inducible through transcriptomic responses in bloom-forming microalgae and through selective breeding.

8.4.1 DETECTION AND MONITORING SYSTEM

Most cultures crash due to infection, contamination, or predation, causing large-scale biomass loss. Recovery can be slow and expensive, leading to significant economic loss. Hence, the primary control strategy is early detection using various methods. Staining and microscopy techniques are traditional methods used to detect the presence of pathogens. Calcofluor white, Nile red, and Congo red are among the most common stains used for this purpose. In addition, FlowCAM (fluid imaging technology) operates by passing samples through a 3 mm × 3 mm glass chamber, where a green laser is illuminated; relative fluorescence data and digital images of particles are then captured and identified. Recently, the FlowCAM technique has been used to monitor microalgal pathogens for early detection (Day et al., 2012). Molecular detection tools have also been developed and represent modern methods for pathogen detection. Molecular methods identify the presence of pathogens more specifically than staining and microscopy techniques. Molecular detection includes qPCR, hybridization (phylochips), and next-generation sequencing (Botes et al., 2013). Spectroscopy, including hyperspectral imaging or reflectance analysis, is used to monitor chlorophyll or other pigment content in ponds, as these parameters change during contamination events. Mass spectrometry is also used to identify different types of contaminants.

8.4.2 SOLUTION AND CONTROL SYSTEM

Salvage harvest is the most basic form of intervention. It is the least used method, as it could also be successful only if there is early detection of the pathogen. This method involves harvesting the ponds before losses occur and a clash happens. Other methods have been published that use chemical agents, such as abscisic acid, fungicides, or changes in environmental status, such as a change in pH or ammonium exposure, which promotes the growth of microalgae and prevents pest development. Physical methods have also been used to kill pathogens without harming microalgae, such as sonication and filtration. Polycultures are biological methods that prevent ponds from contamination (Larkum et al., 2012; Pouneva, 2006).

8.5 IMPACTS OF ENVIRONMENTAL POLLUTANTS ON MICROALGAE

Environmental pollution has been observed since the 19th century due to industrial activities, and certain organic compounds have intensified since then. Many environmental pollutants originate from diverse anthropogenic sources and are capable of degrading and destroying the ecological integrity of microalgae. Marine microalgae, in particular, are promising indicators for both organic and inorganic pollutants since they occupy the base of the food chain and are abundant life forms in the aquatic environment. In remote areas of the Earth, it has been confirmed that in the presence of pesticides, large-scale dispersion and deposition at sites far from the original site of application are observed, and this kind of transport of pollutants occurs in the atmosphere. Wastes produced from various sources, such as domestic sources, industrial activities, and habitat destruction, have a substantial impact on the marine coastal environment. Adverse effects on microalgae include organochlorine compounds, municipal wastes, heavy metals, petroleum products, and domestic herbicides when released at low levels. Bioaccumulative pollutants and toxins are found in trace amounts in water and are often found in sediments. However, data derived from sediments do not represent the exact pollutant concentration accumulated in the overlying water column; hence, it is difficult to produce information based on the patterns of contamination at a higher level of the food chain. For example, the first step of bioaccumulation in the aquatic food web is the uptake of xenobiotics by phytoplankton. The toxic effects of pollutants on microalgae could lead to inhibition of growth and photosynthesis in the microalgae (Burrows, 1971; Torres et al., 2008; Ying and Kookana, 2007).

8.5.1 GENERAL IMPACTS FROM THE ECOSYSTEM

The accumulation of pollutants with low octanol–water partition coefficients (log Kow < 5) occurs in aquatic organisms, while more hydrophobic pollutants with log Kow > 5 may partition into lipid membranes, leading to biomagnification (Islam and Tanaka, 2004).

8.5.1.1 OIL DERIVATIVES

The oil pollution in the ocean is due to fuels accounting for 48% and crude oil for 29% annually. A small portion of the pollution is released by accidents occurring in a few industries and other factors, such as pollutants released from refiners, continuous operational discharges from ships, and tanker collisions. Only 5% of all the pollutants entering the sea are contributed by tanker accidents (Tolosa et al., 2005; Wake, 2005).

8.5.1.2 PESTICIDES

Some pesticides do not undergo rapid degradation; hence, it is likely that a large volume of pesticide residues accumulates in the environment. The pesticides have a wide range of factors that are applied to agricultural soil: active ingredients, method of application, soil type, weather conditions, and so forth. Therefore, extensive contamination of non-target sites and the overuse of these kinds of pesticides will lead to negative impacts on non-target sites, which include aquatic biota, terrestrial plants, and microalgae (Bromilow et al., 2006; Islam and Tanaka, 2004).

8.5.1.3 PERSISTENT ORGANIC POLLUTANTS (POPS)

POPs are also known as hydrophobic organic compounds (HOCs). POPs contain a group of organic chemicals that are bound to bromide or chlorine atoms. The secondary products or residues of agrochemical industries are part of some POPs, such as polychlorinated dibenzo-p-dioxins. The distribution and exchange dynamics of POPs differ by physical phases, that is, in a gas phase or attached to airborne particles, which signifies the difficulty in tracing the movements of these compounds in the environment (Haynes and Johnson, 2000).

8.5.1.4 POLYCYCLIC AROMATIC HYDROCARBONS (PAHS)

PAHs are among the most bio-persistent and biologically toxic components, accounting for 20% of crude oil. These include a range of compounds that have two or more condensed aromatic rings, either with or without alkyl group substituents. These compounds are rapidly absorbed by particulate matter and living organisms; this is due to their low vapor pressures and log Kow > 5. The Environmental Protection Agency (EPA) has listed 16 PAHs as priority pollutants, and some of these are known to be human carcinogens and among the most aggressive pollutants, such as benzo[a]anthracene, chrysene, benzo[b]fluoranthene, benzo[a]pyrene, indeno[1,2,3-cd]pyrene, dibenzo[a,h]anthracene, and benzo[g,h,i]perylene (Boehm et al., 2004).

8.6 MICROALGAE AS AN ECOTOXICITY INDICATOR

In the aquatic system, a large amount of pesticides is found, which run off from the agriculture industry, and as a consequence, organic pollutants appear in the aquatic ecosystem. Microalgae, being the primary producers of the ecosystem due to their microscopic size, constitute the first level of contamination, and because of their short generation time, the effect of pollutants on microalgae will rapidly affect the higher trophic levels of the food chain, leading to significant alterations in the biological integrity of the ecosystem. The oldest way of analyzing the quality of a water stream is based on the indicator species concept, also known as the saprobien system. This kind of system is widely used for monitoring wastewater and municipal water, which allows differentiation between clean and polluted water. As technology improves, it has become possible to observe microalgae due to their microscopic size and to obtain samples at the community and population levels. Microalgal species can be easily cultivated in photobioreactors (PBRs) under adequately controlled conditions (Nie et al., 2009). Transgenerational effects of pollution exposure in microalgae have been researched and studied as long-term pollutant exposure at different population levels.

In the aquatic environment, microalgae can absorb pollutants, bio transform organic compounds, and immobilize inorganic elements to make them less toxic. They are well known to be the basis of pollutant biomagnification and transfer to upper levels of the food web. It has been observed that the presence of pollutants can induce nitrosative and oxidative stress. Since both microalgae and macroalgae have a very important antioxidant system, they are used as powerful biomarker tools for pollution exposure. Microalgae are bioindicators of ecosystem conditions. They can serve as indicators of pollution through a few morphological and chemical characteristics. For example, changes in the pH level of water due to water acidity, discharge of heavy metals from industries, and the maximum effect on the microalgal population can indicate pollution. Due to this, microalgae could be classified by their level of tolerance to certain pollutants. It could be stated that certain species of microalgae could be defined in various zones of degradation in the river, which could be used to indicate the amount of pollution present in the river. For the study of water quality, a numerical approach has been done using diatom flora, which acts as a powerful indicator of environmental changes and monitoring studies. Various types of river degradation are known through the use of algal parameters (Munir et al., 2013).

Inorganic pollution occurs from hazardous chemical fertilizers used for agriculture, detergents used for laundry, and stormwater runoff. Organic pollution occurs from sewage-related sources, which include livestock waste and leaky septic tanks. Lakes and rivers that receive these sources of pollution chronically and periodically indicate high damage to algal growth, resulting in toxin-producing genera or blooms of a nuisance (Carpenter et al., 1924). Toxin-producing blooms can be responsible for hypoxic conditions and may disturb the food web, affecting zooplankton, small fish, large fish, and birds. Toxic blooms can cause problems for households that receive their drinking water from sources like lakes and reservoirs. Toxic blooms restrict recreation, such as fishing, pet-related activities, and swimming. A few microalgae are associated with municipal sewage treatment and are present in very large densities in sewage stabilization ponds, also known as lagoons. These microalgae, which belong to this group in organically polluted water, are rich in phosphorus and nitrogen and are used as biological indicators for organic pollution. The assigned pollution index value ranges from 1 to 6. A density of 50 or more individuals in a 1 ml sample is found to be the index value by microscopic analysis. A value of high organic pollution in the water body is noted as more than 20, and scores varying from 15 to 19 indicate slight organic pollution. Certain green microalgae, diatoms, and flagellate populations and species diversity reflect different trophic conditions, which are indeed important indicators for evaluating water quality.

Organisms may also be exposed to high levels of toxic agents upon the outfall of pollutants in coastal waters. The xenobiotics in the aquatic ecosystem can partition between land, sediment-water interface, biota, sediment, and air-water interface. Therefore, the dynamic behavior of pollutants exposed to the environment is under the influence of atmospheric conditions, water, and abiotic and biotic materials. Microalgae interacting with pollutants present in the atmosphere alter the growth and development of the organisms; they respond differently to different kinds of pollutants, and there are adverse effects on the environment itself. Extensive research has been documented worldwide regarding the impact of organic pollutants on microalgae in soil and the aquatic environment.

The non-taxonomic measurements and the biotic indices reflect the condition of water quality, but these methods are only partially reliable and could be considered descriptive of algal indicators because no specific group of organisms is identified for assessing and detecting disturbances in the environment associated with various human activities. Indicators that fall under a certain group must be included in water quality monitoring analyses to provide an accurate signal of ecosystem change. Scientists have proposed using oxidative stress as a biomarker in microalgae as an ecotoxicity indicator. In a normal, non-polluted environment, microalgae produce free radicals. These cells, as a result of vital activity, are involved in many biological processes and are normally neutralized by antioxidants. Oxidative stress acts as a self-defense mechanism. If the first attempt of the self-defense mechanism fails, the number of free radicals increases, as does the number of antioxidants. However, if disorganization occurs when ROS and RNS (reactive nitrogen species) levels increase, this will lead to damage to the cell's DNA, lipids, and proteins and cause cell death. The biochemical level of ecotoxicity is identified at the earliest stages. Even a small impact on photosynthesis involves many biochemical pathways that depend on the constant flow of electrons. This kind of change will lead to an increase in ROS and protective antioxidants. The efficiency of photosynthesis changes, and the pigment content changes, leading to a significant disturbance in the growth rate. Therefore, by observing changes in the activity of biochemical indicators, one can indicate the presence of hazardous or toxic pollutants. Figure 8.5 shows the principle of using microalgae antioxidants as xenobiotic biomarkers (Pikula et al., 2019; Rezayian et al., 2019).

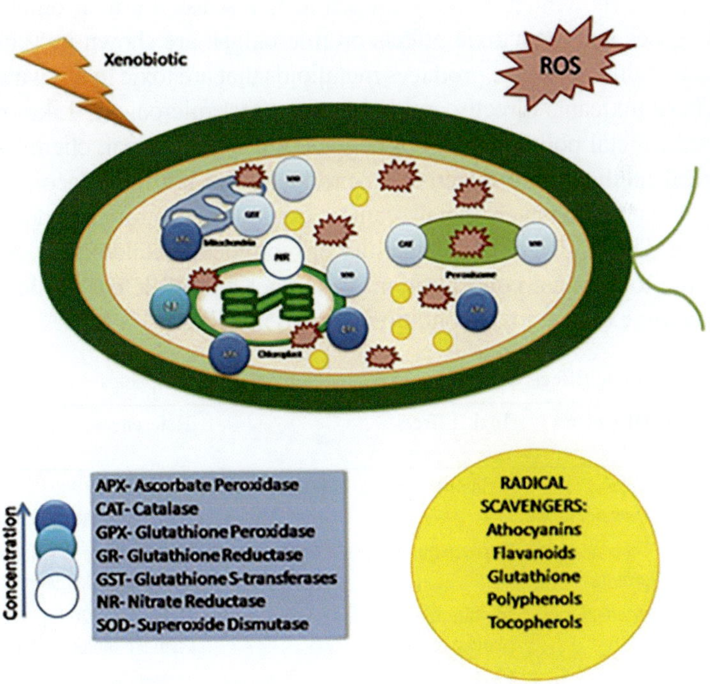

FIGURE 8.5 Microalgal antioxidants acting as stress biomarkers.

8.6.1 TOXICITY FROM ATMOSPHERIC POLLUTANTS AND METAL(LOIDS)

Recently, climatic changes have been mainly caused by CO_2 and CH_4 present in the atmosphere. If carbon dioxide (CO_2) increases by about 3.4 Gt C per year (Houghton et al., 2001), it would affect growth and alter the rates of photosynthesis in microalgae, resulting in carbon storage through the processes of respiration, decomposition, and photosynthesis. Hence, this leads to greenhouse effects. At present, more efforts are being made to increase the utilization and decrease the production of CO_2. A permanent solution could be that microbial and plant groups act as sinks for CO_2. The increasing consequences of global warming are due to increases in UV-B radiation and atmospheric concentrations of oxygen and CO_2, which further alter the major activities of microalgae. The absorption of CO_2 by microalgae is a prominent feature that makes them good bioindicators (Hadiyanto et al., 2012). The cyanobacterium *Anabaena* sp. resulted in oxidative damage involving reactive O_2 species caused by UV-B radiation (He and Hader, 2002). UV effects on algae are crucial, as they cause damage to mitochondria and chloroplasts, which is likely related to UV stress. Further, other atmospheric gases and their toxic effects on microalgae are shown in Table 8.2. Domestic industrial waste produces metalloids that are toxic to air, water, and soil. These toxicants directly and indirectly affect microalgae. Like organic pollutants, metal pollutants do not degrade through physical, chemical, and biological methods into different by-products, nor are they decomposable by any means. However, metals can be transformed through reduction or oxidation. The degree of bioavailability or biological metal toxicity does not reflect the total metal concentration. In the act of self-defense, microalgae can overcome certain toxic conditions shown in Table 8.3.

TABLE 8.2 Toxic Effects of Atmospheric Gases on Microalgal Species

Atmospheric Gases	Organisms	Toxic Effects	References
Oxygen	*Nostoc spongiaeforme*	Oxidative stress at 1 or 2 mM H_2O_2	Hirata et al. (2003); Ramakrishnan et al. (2010)
Carbon dioxide	*Scenedesmus armatus*	Toxic to quinones at 2%	Ramakrishnan et al. (2010); Tukaj and Aksmann (2007)
Benzene hexachloride	*Anabaena* sp.	Inhibition of biomass production and photosynthetic rate	Babu et al. (2001); Ramakrishnan et al. (2010)
Atmospheric methane	*Chaetoceros* sp.	Production of hexadecatrienoic acid, an inhibitor of methanogenesis	Ramakrishnan et al. (2010); Ungerfeld et al. (2005)

TABLE 8.3 Toxic Influence and Mechanisms of Detoxification of Selected Metal(loid)s

Metal(loid)	Toxic Effects	Mechanism of Detoxification	References
Chromium	Cell membrane disruption, DNA damage, protein denaturation, and inhibition of cell division, transcription, translation, and enzymatic activity	Adsorption, precipitation, reduction of Cr(IV) to Cr(III)	Cervantes et al. (2001); Deng et al. (2006)
Copper		Reduction (Cd^{2+} to $CdPO_4$)	Abdale et al. (1995); Cid et al. (1996); Cid et al. (1997); Eisler (1998); Bossuyt and Janssen (2005)
Zinc		Adsorption and sequestration by metallothionein	Cavet et al. (2003)
Arsenic		Adenosylation, reduction, and oxidative methylation	Dembitsky and Levitsky (2004)
Cadmium		Production of metallothionein-like protein ($-(cys\text{-}cys)_n\text{-}Cd$)	Clemens (2006)
Mercury		Volatilization and reduction	Deng et al. (2008)
Lead		Sequestration of intracellular polyphosphate bodies	Battah et al. (2001); Raungsomboon et al. (2006)

8.6.2 MICROALGAE ON TOXICITY STUDIES

The physiological changes in microalgae, such as inhibition of growth and photosynthesis, as well as ATP formation, oxygen evolution, microalgal fluorescence induction, and radioactive carbon assimilation, act as potential indicators of hazardous and toxic effects of pollutants on microalgae. Some of the recent advances to detect early warning signals of sublethal levels of pollution have been published using the flow cytometry technique.

8.6.2.1 CHANGES IN CELL MORPHOLOGY AND CELL PHYSIOLOGY

When microalgae are exposed to toxic pollutants, it mainly induces morphological changes. The ultrastructure or the structure of a microalgal cell is studied using light and electron microscopy techniques by viewing the morphological changes in cell shape and volume, as well as the morphological changes in mitochondria and chloroplasts (Yang et al., 2002). For example, cadmium metal caused ultrastructural changes in *Phaeodactylum tricornutum*: reduction of lipid inclusions, increase in the volume of the chloroplast, and high deposition of metal particles on the cell surface (Torres et al., 2008). The toxic environmental pollutants can also lead to different enzyme activities either by inducing or inhibiting the presence of toxic pollutants. This kind of environmental stress in microalgae is considered and studied more. Generally, the environmental contaminant disturbs the oxidative stress mechanism, which in turn induces antioxidative enzymatic and non-enzymatic responses. The study is based on the enzymatic activities related to antioxidative mechanisms such as peroxidase, reductase, glutathione, and catalase. This technique is typically used as biomarkers of toxicity (Geoffroy et al., 2002). Studies of enzymatic activities also include nitrogen metabolism. The herbicide paraquat affected the enzymes involved in nitrogen assimilation, with nitrite reductase activity leading to a significant alteration of the C/N ratio (Franqueira et al., 1999; Prado et al., 2009). Photosynthesis is also considered one of the physiological parameters and is mostly targeted by pollutants. The difference or inhibition of photosynthesis indicates the presence of pollutants against microalgae. There are both traditional and modern methods to monitor photosynthesis activity. The traditional methods include the assimilation of ^{14}C-radiolabeled CO_2 and monitoring the evolution of oxygen through electrodes. The modern method is used as an effective indicator of pollution by the chlorophyll fluorescence method. When a microalgal cell containing pigment is exposed to light, it produces fluorescence originating from photosystem II and chlorophyll a (Strom et al., 2009).

8.6.2.2 GROWTH AND CELL VIABILITY

The physiological state of a microalgal cell is generally determined by its growth and it is the most studied parameter in the toxicity test. The toxic effects on microalgal growth could lead to major changes in the cell cycle proliferation and microalgal cell division. The growth of an algal cell is monitored by two methods, direct and indirect method. The direct method is by counting cells using the microscopic technique in special chambers or using flow cytometry. The indirect method is estimated by observing the turbidity of the culture or the amount of chlorophyll a present by using techniques like spectrophotometry or fluorometry. The determination of the cell viability of a microalgal cell is performed by detecting the amount of fluorescence of cells stained with "propidium iodide" using flow cytometry or a microscopic technique. As the cell loses its membrane integrity or dies, this allows the fluorescent dye to penetrate inside the cell. When excited by blue light, the fluorochrome fluoresces red (Franqueira et al., 2000).

8.6.2.3 DNA DAMAGE

Some of the toxic agents tend to damage the DNA of the microalgal cell, causing genotoxic effects. There has been a wide range of studies and tests based on biomarkers regarding the genotoxic effects of environmentally hazardous pollutants (Ali and Kumar, 2008), but as this issue has become serious due to uncontrollable changes, recent advances have made it possible to conduct simple and rapid tests for determining genotoxicity and eliminating DNA damage (Akcha et al., 2008). The tests were able to detect various DNA strands that are damaged by inducing physical or chemical agents. The comet assay is characterized by the formation of so-called "comets" after the migration of DNA in the electrophoresis medium due to single-stranded DNA breaks induced by treatment with alkalis or certain genotoxic agents. Comets, being smaller in size, migrate and consist of a head and a tail, which are visible when stained with a fluorophore dye such as ethidium bromide or SYBR Green. DNA damage induced by alkali, oxidative stress, and intercalating agents can be detected using the comet assay process (Erbes et al., 1997; Henderson et al., 1998).

8.6.2.4 BIOCHEMICAL COMPOSITION OF MICROALGAL BIOMASS

In the cellular regulatory process, metabolism is considered the end product, and changes in metabolites ultimately respond to environmental and genetic changes (Jamers et al., 2009). Chlorophylls and carotenoids are the main photosynthetic pigments in a cell and are related to photosynthetic metabolic activity (Couderchet and Vernet, 2003). For example, *C. moewusii* was analyzed for photosynthetic pigments using a spectrophotometric technique that provides bulk measurements. The flow cytometry method is also used (Gonzalez-Barreiro et al., 2004; Prado et al., 2011).

8.7 CONCLUSION

This chapter mainly focuses on microalgae as potential indicators of the environmental toxicity of hazardous pollutants. As we know, microalgae play a very important role in our ecosystem, as they are the primary producers of the food chain. Microalgae also have various applications for human benefit. Apart from natural predators and pathogens present against microalgae, human activities such as agriculture and industrial activities lead to toxic effects; that is, hazardous pesticides produced from agricultural fields and heavy metal wastes produced from industries degrade the ecosystem. Some of the toxic pollutants are nonbiodegradable, while others are biodegradable. Around 60,000 man-made chemical substances are used today, which increases the chance of pollution. All the pollution in the ecosystem—air, water, and soil—either directly or indirectly affects microalgae. Hence, microalgae are also bioindicators of the ecosystem. Morphological or chemical changes in microalgae will determine whether the ecosystem is polluted. Fortunately, many microalgal species are adapted to a few toxins and have also developed detoxification or defense mechanisms over the years.

ACKNOWLEDGMENTS

We acknowledge REVA University for providing research support.

CONFLICT OF INTEREST

The authors declare no conflict of interest.

KEYWORDS

- **food chain**
- **human carcinogens**
- **hydrophobic organic compounds**
- **microalgae**
- **microalgal species**
- **organic pollution**
- **reactive oxygen species**

REFERENCES

Abalde, J., Cid, A., Reiriz, S., Torres, E., & Herrero, C. (1995). Response of the marine microalga *Dunaliella tertiolecta* (Chlorophyceae) to copper toxicity in short-time experiments. *Bulletin of Environmental Contamination and Toxicology, 54*, 317–324.

Akcha, F., Arzul, G., Rousseau, S., & Bardouil, M. (2008). Comet assay in phytoplankton as biomarker of genotoxic effects of environmental pollution. *Marine Environmental Research, 66*, 59–61.

Ali, D., & Kumar, S. (2008). Long-term genotoxic effect of monostrophes in different tissues of freshwater fish *Channa punctatus* (Bloch) using alkaline single-cell gel electrophoresis. *Science of the Total Environment, 405*, 345–350.

Babu, G. S., Hans, R. K., Singh, J., Viswanathan, P. N., & Joshi, P. C. (2001). Effect of lindane on the growth and metabolic activities of cyanobacteria. *Ecotoxicology and Environmental Safety, 48*, 219–221.

Battah, M., Shabana, E. F., Kobbia, L. A., & Eladel, H. M. (2001). Differential effects of thiobencarb toxicity on growth and photosynthesis of *Anabaena variabilis* with changes in phosphate level. *Ecotoxicology and Environmental Safety, 49*, 235–239.

Berry, J. P., Gantar, M., Gibbs, P. D., & Schmale, M. C. (2007). The zebrafish (*Danio rerio*) embryo as a model system for identification and characterization of developmental toxins from marine and freshwater microalgae. *Comparative Biochemistry and Physiology Part C: Toxicology and Pharmacology, 145*(1), 61–72.

Boehm, P. D., Page, D. S., Brown, J. S., Neff, J. M., & Burns, W. A. (2004). Polycyclic aromatic hydrocarbon levels in mussels from Prince William Sound, Alaska, USA, document the return to baseline conditions. *Environmental Toxicology and Chemistry, 23*(12), 2916–2929.

Bossuyt, B. T. A., & Janssen, C. R. (2004). Long-term acclimation of *Pseudokirchneriella subcapitata* (Korshikov) Hindák to different copper concentrations: Changes in tolerance and physiology. *Aquatic Toxicology, 68*, 61–74.

Bossuyt, B. T. A., & Janssen, C. R. (2005). Copper regulation and homeostasis of *Daphnia magna* and *Pseudokirchneriella subcapitata*: Influence of acclimation. *Environmental Pollution, 136*, 135–144.

Botes, M., de Kwaadsteniet, M., & Cloete, T. E. (2013). Application of quantitative PCR for the detection of microorganisms in water. *Analytical and Bioanalytical Chemistry, 405*(1), 91–108.

Bromilow, R. H., de Carvalho, R. F., Evans, A. A., & Nicholls, P. H. (2006). Behavior of pesticides in sediment/water systems in outdoor mesocosms. *Journal of Environmental Science and Health, Part B, 41*(1), 1–16.

Burrows, E. M. (1971). Assessment of pollution effects by the use of algae. *Proceedings of the Royal Society of London. Series B, 177*, 295–306.

Carney, L. T., & Lane, T. W. (2014). Parasites in algae mass culture. *Frontiers in Microbiology, 5*, 278.

Carpenter, K. E. (1924). A study of the fauna of rivers polluted by lead mining in the Aberystwyth district of Cardiganshire. *Annals of Applied Biology, 11*, 1–23.

Cavet, J. S., Borrelly, G. P. M., & Robinson, N. J. (2003). Zn, Cu and Co in cyanobacteria: Selective control of metal availability. *FEMS Microbiology Reviews, 27*, 165–181.

Cervantes, C., Campos-García, J., Devars, S., Gutierrez-Corona, F., Loza-Tavera, H., Torres-Guzmán, J. C., et al. (2001). Interactions of chromium with microorganisms and plants. *FEMS Microbiology Reviews, 25*, 335–347.

Chu, W. L. (2012). Biotechnological applications of microalgae. *IeJSME, 6*(1), S24–S37.

Cid, A., Fidalgo, P., Herrero, C., & Abalde, J. (1996). Toxic action of copper on the membrane system of a marine diatom measured by flow cytometry. *Cytometry, 25*, 32–36.

Cid, A., Herrero, C., Torres, E., & Abalde, J. (1995). Copper toxicity on the marine microalga *Phaeodactylum tricornutum*: Effects on photosynthesis and related parameters. *Aquatic Toxicology, 31*, 165–174.

Cid, A., Torres, E., Herrero, C., & Abalde, J. (1997). Disorders provoked by copper in the marine diatom *Phaeodactylum tricornutum* in short-time exposure assays. *Cahiers de Biology Marine, 38*, 201–206.

Clemens, S. (2006). Toxic metal accumulation, responses to exposure and mechanisms of tolerance in plants. *Bioclimate, 88*, 1707–1719.

Couderchet, M., & Vernet, G. (2003). Pigments as biomarkers of exposure to the vineyard herbicide flazasulfuron in freshwater algae. *Ecotoxicology and Environmental Safety, 55*, 271–277.

Day, J. G., Thomas, N. J., Achilles-Day, U. E., & Leakey, R. J. (2012). Early detection of protozoan grazers in algal biofuel cultures. *Bioresource Technology, 114*, 715–719.

Dazy, M., Masfaraud, J. F., & Férard, J. F. (2009). Induction of oxidative stress biomarkers associated with heavy metal stress in *Fontinalis antipyretica* Hedw. *Chemosphere, 75*(3), 297–302.

Dembitsky, V. M., & Levitsky, D. O. (2004). Arsen lipids. *Progress in Lipid Research, 43*, 403–448.

Deng, L., Wang, H., & Deng, N. (2006). Photoreduction of chromium(VI) in the presence of alga *Chlorella vulgaris*. *Journal of Hazardous Materials, 138*, 288–292.

Deng, L., Wu, F., Deng, N., & Zuo, Y. (2008). Photoreduction of mercury (II) in the presence of algae, *Anabaena cylindrica*. *Journal of Photochemistry and Photobiology B: Biology, 91*, 117–124.

Eisler, R. (1998). *Copper Hazards to Fish, Wildlife, and Invertebrates: A Synoptic Review* (No. 33). U.S. Geological Survey.

Enamala, M. K., Enamala, S., Chavali, M., Donepudi, J., Yadavalli, R., Kolapalli, B., et al. (2018). Production of biofuels from microalgae—A review on cultivation, harvesting, lipid extraction, and numerous applications of microalgae. *Renewable and Sustainable Energy Reviews, 94*, 49–68.

Erbes, M., Webles, A., Obst, U., & Wild, A. (1997). Detection of primary DNA damage in *Chlamydomonas reinhardtii* by means of modified microgel electrophoresis. *Environmental and Molecular Mutagenesis, 30*, 448–458.

Eullaffroy, P., & Vernet, G. (2003). The F684/F735 chlorophyll fluorescence ratio: A potential tool for rapid detection and determination of herbicide phytotoxicity in algae. *Water Research, 37*, 1983–1990.

Franklin, N. M., Adams, M. S., Stauber, J. L., & Lim, R. P. (2001). Development of an improved rapid enzyme inhibition bioassay with marine and freshwater microalgae using flow cytometry. *Archives of Environmental Contamination and Toxicology, 40*(4), 469–480.

Franqueira, D., Cid, A., Torres, E., Orosa, M., & Herrero, C. (1999). A comparison of the relative sensitivity of structural and functional cellular responses in the alga *Chlamydomonas eugametos* exposed to the herbicide paraquat. *Archives of Environmental Contamination and Toxicology, 36*, 264–269.

Franqueira, D., Orosa, M., Torres, E., Herrero, C., & Cid, A. (2000). Potential use of flow cytometry in toxicity studies with microalgae. *Science of the Total Environment, 247*, 119–126.

Gachon, C. M. M., Sime-Ngando, T., Strittmatter, M., Chambouvet, A., & Kim, G. H. (2010). Algal diseases: Spotlight on a black box. *Trends in Plant Science, 15*(11), 633–640.

Geoffroy, L., Teisseire, H., Couderchet, M., & Vernet, G. (2002). Effect of oxyfluorfen and diuron alone and in mixture on antioxidative enzymes of *Scenedesmus obliquus*. *Pesticide Biochemistry and Physiology, 72*, 178–185.

Gonzalez-Barreiro, O., Rioboo, C., Cid, A., & Herrero, C. (2004). Atrazine-induced chlorosis in *Synechococcus elongatus* cells. *Archives of Environmental Contamination and Toxicology, 46*, 301–307.

Grami, B., Rasconi, S., Niquil, N., Jobard, M., Saint-Béat, B., & Sime-Ngando, T. (2011). Functional effects of parasites on food web properties during the spring diatom bloom in Lake Pavin: A linear inverse modeling analysis. *PLoS ONE, 6*(8), e23273.

Hadiyanto, H. (2012). Bio fixation of carbon dioxide by *Chlamydomonas* sp. in a tubular photobioreactor. *International Journal of Renewable Energy Development, 1*(1), 10–14.

Hallmann, A. (2007). Algal transgenics and biotechnology. *Transgenic Plant Journal, 1*(1), 81–98.

Haynes, D., & Johnson, J. E. (2000). Organochlorine, heavy metal and polyaromatic hydrocarbon pollutant concentrations in the Great Barrier Reef (Australia) environment: A review. *Marine Pollution Bulletin, 41*(7–12), 267–278.

He, Y. Y., & Häder, D. P. (2002). UV-B-induced formation of reactive oxygen species and oxidative damage of the cyanobacterium *Anabaena* sp.: Protective effects of ascorbic acid and N-acetyl-L-cysteine. *Journal of Photochemistry and Photobiology B: Biology, 66*(2), 115–124.

Henderson, L., Wolfreys, A., Fedyk, J., Bourner, G., & Windebank, S. (1998). The ability of the comet assay to discriminate between genotoxins and cytotoxins. *Mutagenesis, 13*, 89–94.

Hirata, K., Yoshitomi, S., Dwi, S., Iwabe, O., Mahakhant, A., Polchai, J., & Miyamoto, K. (2003). Bioactivities of nostocine A produced by a freshwater cyanobacterium *Nostoc spongiaeforme* TISTR 8169. *Journal of Bioscience and Bioengineering, 95*, 512–517.

Houghton, R. A., Hackler, J. L., & Cushman, R. M. (2001). *Carbon Flux to the Atmosphere from Land-Use Changes: 1850 to 1990*. Oak Ridge National Laboratory.

Islam, M. S., & Tanaka, M. (2004). Impacts of pollution on coastal and marine ecosystems including coastal and marine fisheries and approach for management: A review and synthesis. *Marine Pollution Bulletin, 48*(7, 8), 624–649.

Jamers, A., Blust, R., & De Coen, W. (2009). Omics in algae: Paving the way for a systems biological understanding of algal stress phenomena? *Aquatic Toxicology, 92*(3), 114–121.

Kellam, S. J., Cannell, R. J. P., Owsianka, A. M., & Walker, J. M. (1988). Results of a large-scale screening programmed to detect antifungal activity from marine and freshwater microalgae in laboratory culture. *British Phycological Journal, 23*(1), 45–47.

Kumar, M. S., & Buddolla, V. (2019). Future prospects of biodiesel production by microalgae: A short review. In *Recent Developments in Applied Microbiology and Biochemistry* (pp. 161–166). Academic Press.

Larkum, A. W. D., Ross, I. L., Kruse, O., & Hankamer, B. (2012). Selection, breeding and engineering of microalgae for bioenergy and biofuel production. *Trends in Biotechnology, 30*(4), 198–205.

Munir, N., Sharif, N., Naz, S., & Manzoor, F. (2013). Algae: A potent antioxidant source. *Sky Journal of Microbiology Research, 1*(3), 22–31.

Nie, J., & Sage, E. H. (2009). SPARC functions as an inhibitor of adipogenesis. *Journal of Cell Communication and Signaling, 3*(3, 4), 247.

Pikula, K. S., Zakharenko, A. M., Aruoja, V., Golokhvast, K. S., & Tsatsakis, A. M. (2019). Oxidative stress and its biomarkers in microalgal ecotoxicology. *Current Opinion in Toxicology, 13*, 8–15.

Pouneva, I. D. (2006). Effect of abscisic acid and oncogenic phases of the host alga on the infection process in the path system *Scenedesmus acutus—Phlyctidium scenenedesmi. Acta Physiologiae Plantarum, 28*(5), 395–400.

Prado, R., García, R., Rioboo, C., Herrero, C., Abalde, J., & Cid, A. (2009). Comparison of the sensitivity of different toxicity test endpoints in a microalga exposed to the herbicide paraquat. *Environment International, 35*(2), 240–247.

Prado, R., Rioboo, C., Herrero, C., & Cid, Á. (2011). Characterization of cell response in *Chlamydomonas moewusii* cultures exposed to the herbicide paraquat: Induction of chlorosis. *Aquatic Toxicology, 102*(1, 2), 10–17.

Priyadarshani, I., & Rath, B. (2012). Commercial and industrial applications of microalgae—A review. *Journal of Algal Biomass Utilization, 3*(4), 89–100.

Ramakrishnan, B., Megharaj, M., Venkateswarlu, K., Naidu, R., & Sethunathan, N. (2010). The impacts of environmental pollutants on microalgae and cyanobacteria. *Critical Reviews in Environmental Science and Technology, 40*(8), 699–821.

Raungsomboon, S., Chidthaisong, A., Bunnag, B., Inthorn, D., & Harvey, N. W. (2008). Removal of lead (Pb^{2+}) by the cyanobacterium *Gloeocapsa* sp. *Bioresource Technology, 99*, 5650–5658.

Ray, M., Kumar, N., Kumar, V., Negi, S., & Banerjee, C. (2019). Microalgae: A way forward approach towards wastewater treatment and bio-fuel production. In *Applied Microbiology and Bioengineering* (pp. 229–243). Academic Press.

Rezayian, M., Niknam, V., & Ebrahimzadeh, H. (2019). Oxidative damage and antioxidative system in algae. *Toxicology Reports, 6*, 1309–1313.

Sigamani, S., Ramamurthy, D., & Natarajan, H. (2016). A review on potential biotechnological applications of microalgae. *Journal of Applied Pharmaceutical Science, 6*(8), 179–184.

Simkovsky, R., Daniels, E. F., Tang, K., Huynh, S. C., Golden, S. S., & Brahamsha, B. (2012). Impairment of O-antigen production confers resistance to grazing in a model amoeba–cyanobacterium predator–prey system. *Proceedings of the National Academy of Sciences of the United States of America, 109*(41), 16678–16683.

Sousa, I., Gouveia, L., Batista, A. P., Raymundo, A., & Bandarra, N. M. (2008). Microalgae in novel food products. *Food Chemistry Research Developments*, 75–112.

Spolaore, P., Joannis-Cassan, C., Duran, E., & Isambert, A. (2006). Commercial applications of microalgae. *Journal of Bioscience and Bioengineering, 101*(2), 87–96.

Strom, D., Ralph, P. J., & Stauber, J. L. (2009). Development of a toxicity identification evaluation protocol using chlorophyll-a fluorescence in a marine microalga. *Archives of Environmental Contamination and Toxicology, 56*(1), 30–38.

Talukder, M. M. R., Das, P., & Wu, J. C. (2014). Immobilization of microalgae on exogenous fungal mycelium: A promising separation method to harvest both marine and freshwater microalgae. *Biochemical Engineering Journal, 91*, 53–57.

Tolosa, I., De Mora, S. J., Fowler, S. W., Villeneuve, J. P., Bartocci, J., & Cattini, C. (2005). Aliphatic and aromatic hydrocarbons in marine biota and coastal sediments from the Gulf and the Gulf of Oman. *Marine Pollution Bulletin, 50*(12), 1619–1633.

Torres, M. A., Barros, M. P., Campos, S. C., Pinto, E., Rajamani, S., Sayre, R. T., et al. (2008). Biochemical biomarkers in algae and marine pollution: A review. *Ecotoxicology and Environmental Safety, 71*(1), 1–15.

Tukaj, Z., & Aksmann, A. (2007). Toxic effects of anthraquinone and phenanthrenequinone upon *Scenedesmus* strains (green algae) at low and elevated concentration of CO_2. *Chemosphere, 66*, 480–487.

Ungerfeld, E. M., Rust, S. R., Burnett, R. J., Yokoyama, M. T., & Wang, J. K. (2005). Effects of two lipids on *in vitro* ruminal methane production. *Animal Feed Science and Technology, 119*, 179.

Vandamme, D., Pontes, S. C. V., Goiris, K., Foubert, I., Pinoy, L. J. J., & Muylaert, K. (2011). Evaluation of electro-coagulation–flocculation for harvesting marine and freshwater microalgae. *Biotechnology and Bioengineering, 108*(10), 2320–2329.

Wake, H. (2005). Oil refineries: A review of their ecological impacts on the aquatic environment. *Estuarine, Coastal and Shelf Science, 62*(1, 2), 131–140.

Yang, S., Wu, R. S., & Kong, R. Y. (2002). Biodegradation and enzymatic responses in the marine diatom *Skeletonema costatum* upon exposure to 2,4-dichlorophenol. *Aquatic Toxicology, 59*(3, 4), 191–200.

Ying, G. G., & Kookana, R. S. (2007). Triclosan in wastewaters and biosolids from Australian wastewater treatment plants. *Environment International, 33*(2), 199–205.

CHAPTER 9

Environmental Flue Gas Cleanup by Green Algae and Cyanobacteria

BURCU ERTIT TAŞTAN[1,2,*]

[1]*Vocational School of Health Services, Gazi University, Ankara, Türkiye*

[2]*Life Sciences Application and Research Center, Gazi University, Ankara, Türkiye*

*Corresponding author

ABSTRACT

The rise in greenhouse gases due to industrialization and human activities is the main cause of global climate change. The accelerating use of fossil fuels due to global industrialization has exponentially increased environmental pollution. Changes in Earth's climate system and global warming are felt everywhere in the world, from the equator to the poles. This chapter reports on the concept of algal cultivation techniques for current and future environmental cleanup. In this chapter, the advantages and challenges of algal cultivation techniques by means of a carbon capture strategy are discussed. Microalgae are very important microorganisms that have the potential to solve environmental and energy problems in the world. These photosynthetic organisms perform the synthesis of many structural materials as well as convert CO_2 to O_2 with the energy they receive from the sun. These microorganisms, which represent the most natural and cheapest method of CO_2 removal, may be among the biofuels of the future because they reduce dependence on fossil resources.

Environmental Phycology: Climate Change Mitigation and Hazardous Pollutant Cleanup Using Algal Resources. Jeyabalan Sangeetha and Devarajan Thangadurai (Eds.)

DOI: 10.1201/9781779647085-9

9.1 INTRODUCTION

The endless increase in greenhouse gas emissions has led to changes in Earth's climate system, and global warming is evident everywhere in the world, from the equator to the poles. The noticeable effects of temperature increases on the ecosystem have now reached an irreversible point over the last 50 years. Anthropogenic activities have led to floods, while in some regions severe droughts and desertification are prevalent. The rise in winter temperatures, early springs, the decrease in plant and animal species, the increase in extinct species, and all effects of climate change are turning the ecosystem upside down. The changing climate system and increasing greenhouse gas emissions are dragging the universe toward extinction step by step unless effective action is taken immediately.

The increase in energy demand and CO_2 emissions continues to rise excessively, and fossil fuels continue to trend as the main energy source. Fossil fuels are the main energy sources worldwide, although renewable energy production is increasing rapidly. While demand for oil, natural gas, and coal is expected to increase by 2035, it is emphasized that the share of the energy composition will decrease from 81% to 75% (IEA, 2012). Coal remains the backbone of global electricity generation, especially in non-OECD countries. In addition, increasing vehicle traffic and heavy pollution from vehicles threaten the universe, especially urban areas.

According to the International Energy Agency report, global energy demand is expected to increase by at least one-third by 2035. Energy-induced CO_2 emissions are expected to increase from 31.2 Gt in 2011 to 37.0 Gt in 2035, and the global temperature is expected to increase by an average of 3.6C in the long term (IEA, 2012). The main factor in increasing greenhouse gas emissions and climate change is the rise in the CO_2 level in the atmosphere. CO_2 emissions were measured at 399.72 ppm in April 2013 (IEA, 2012). According to scientists' estimates, exceeding the 400 ppm level means exceeding the critical threshold at which average temperatures reach dangerous levels.

CO_2 emissions, known as the most important greenhouse gases, have become increasingly significant over the past 25 years. The amount of CO_2 emitted from 5.47 kg of gasoline per 100 km of travel is calculated as 21.1 kg of CO_2 (Matsumoto et al., 1997a). Many countries have signed the Kyoto Protocol, the international agreement to combat global warming and climate change, to reduce CO_2 emissions worldwide. The Kyoto Protocol is an international protocol to combat global warming and climate change.

9.1.1 AIR POLLUTION

Air is one of the most important factors for organisms to survive. The air volume a person needs daily is between 10 m^3 and 20 m^3. Humans can survive 60 days without food and six days without water but can survive only six minutes without air (RSHC, 2004). Air pollution occurs as a result of changes in the amount of compounds in its natural structure and the addition of foreign materials to this structure.

9.1.2 COMPOSITION OF THE ATMOSPHERE

According to the report published in 2004 by the Environmental Health Research Directorate of Refik Saydam Hygiene Center, Türkiye, the composition of the atmosphere is classified as follows (RSHC, 2004):

- Gases that are constantly present in the air and whose quantities do not change (nitrogen, oxygen, inert gases);
- Gases that are constantly present in the air and whose quantities decrease and increase (carbon dioxide, water vapor, ozone);
- Gases not always present in the air (pollutants).

The volumetric ratios of gases in the atmosphere up to 25 km above ground remain stable due to horizontal and vertical air movements. However, humidity and CO_2 show certain variability (Machida et al., 2003). Due to the absence of vertical air movements in the upper layers, the gases are arranged in layers according to Dalton's law in accordance with their weight. The amounts of gases up to 25 km in the atmosphere are specified in Table 9.1 (RSHC, 2004).

TABLE 9.1 Amounts of Gases up to 25 km in the Atmosphere

Gas	Volume (%)
N_2	78.08
O_2	20.94
Ar	0.93
CO_2	0.03
Ne	0.0018
He	0.0005
H_2O	0.004

When taking into consideration the criteria of energy efficiency, environmental and potential impacts, it is essential to implement measures that are not regional but universal, that have the capacity to act in the long term, that are renewable and sustainable, and that avoid secondary environmental damage.

In this context, microalgae are very important microorganisms that have the potential to solve environmental and energy problems in the world (de-Bashan and Bashan, 2010). Until microalgae appeared 3 million years ago and converted CO_2 to O_2 through photosynthesis, our planet was not suitable for life due to excessive amounts of N_2, CO_2, and toxic gases in the atmosphere. Subsequently, the Sun's rays striking O_2 molecules formed by microalgae led to the recombination into O_3 and the formation of the ozone layer that protects the earth's surface (Vassy, 1965). After all, we would not exist without microalgae. Today, we use the potential of microalgae to save our planet from global warming and greenhouse gas effects (Van Den Hende, 2011, 2012).

As is known, CO_2 is converted into biomass through the metabolism of microalgae. These photosynthetic organisms perform the synthesis of many structural materials as well as convert CO_2 to O_2 using the energy they receive from the sun. These microorganisms, which represent the most natural and cheapest method of CO_2 removal, may be among the biofuels of the future because they reduce dependence on fossil resources.

9.2 MICROALGAE

Microalgae have two basic cell types: prokaryotic and eukaryotic. Prokaryotic cells are cells without a membrane-bound nucleus and are found only in the Cyanobacteria group. They are called blue-green algae according to their color in the culture media (Figure 9.1).

Eukaryotic cells are surrounded by a cell wall consisting of polysaccharides and secreted by the Golgi body. This structure includes the plasma membrane, a living structure as mentioned in the sources, and contains a nucleus. Chloroplasts have membrane layers called thylakoids, where the light reactions of photosynthesis occur. Chloroplasts are surrounded by a double-layered chloroplast membrane. The thylakoids are embedded in the stroma, and the dark reactions of photosynthesis occur in the stroma. One of the best-known groups of eukaryotic microalgae is green algae (Figure 9.2).

FIGURE 9.1 Light microscopy, cultivation, and biomass images of blue-green algae.
Source: Courtesy of Burcu Ertit Taştan, Gazi University, Ankara, Türkiye (2020).

FIGURE 9.2 Confocal microscopy, cultivation, and biomass images of blue-green algae.
Source: Courtesy of Burcu Ertit Taştan, Gazi University, Ankara, Türkiye (2020).

9.2.1 ECONOMIC IMPORTANCE OF MICROALGAE

The algae cultivated in both natural and laboratory conditions are of great economic importance. This importance stems from the widespread use of algae in many research fields, and they have been particularly popular for many years. The main application areas of microalgae can be grouped as environmental biotechnology, bioenergy production, food, food supplements, and the synthesis of certain chemicals (Becker, 2008; de Bashan and Bashan, 2010). The key factor for successful microalgae production is "C," which is the main element of microalgae. It constitutes 36%–65% of the dry weight of microalgae (Chae et al., 2006; Syndey et al., 2010).

Today, microalgae are used in a wide range of applications, from wastewater treatment to solar energy and biomass production; from improving air quality by removing excess CO_2 from the environment to the control of water quality; and from the production of certain chemicals to energy recovery. In addition, due to their high content of carbohydrates, protein, and especially fatty acids, they are used in animal feed, food additives, cosmetics, and medicine. They are an important food source for aquatic organisms.

9.3 PERSPECTIVES OF MICROALGAL CULTIVATION

The sun is the universal source of life energy. Photosynthesis is the basic energy-supplying strategy for algae. Photosynthesis reactions are divided into two groups: light-dependent reactions and light-independent reactions. Light-dependent reactions require the capture of light energy and its conversion into ATP and NADPH. Light-independent reactions involve a series of reactions in which this chemical energy is used to fix inorganic carbon and reduce it. On the basis of their nutritional requirements, algae can be classified as (Barsanti and Gualteri, 2014): (i) obligate heterotrophic algae; (ii) obligate phototrophic algae; (iii) facultative mixotrophic algae; and (iv) obligate mixotrophic algae.

Photoautotrophic microalgae use sunlight as the source of energy and CO_2 as the source of carbon to produce carbohydrates and metabolic energy. On the other hand, some microalgae use a complex nutritional strategy combining photoautotrophy and heterotrophy. This type of nutritional strategy is called mixotrophy. Some mixotrophs are photosynthetic, whereas others meet most of their nutritional requirements through phagotrophy. Heterotrophy becomes important when light is limited.

Microalgae can develop in almost any habitat and nearly anywhere in the world. The culture media and culture conditions should resemble the alga's natural environment as closely as possible. Sterile cultures of microalgae are maintained as unialgal, contaminant-free, and axenic cultures. To maintain a successful microalgae culture, several distinct parameters are required. Cultivation should include a suitable vessel, a suitable species-specific medium, and a means to allow CO_2 exchange between the medium and the atmosphere. Other important factors are light, water, and nutrients. In addition, optimal parameters such as temperature, pH, and salinity are necessary for the successful growth of microalgae.

Choosing the correct culture medium is the best way to maintain algal cultures. The natural habitats of microalgae should be taken into consideration when preparing a culture medium. It is also important to know the nutritional strategy of the cultured microalgae, whether mixotrophic, heterotrophic, or phototrophic. There are a number of media recipes, such as freshwater or marine media. BG11 medium (Rippka, 1988), Bold's basal medium (Dayananda et al., 2007), MES-Volvox medium, Zarrouk's medium, Jaworski's medium, CHU medium, f/2 medium, and Johnson's medium (Barsanti and Gualteri, 2014) are commonly used media types.

Microalgae can be cultured using a wide variety of methods, from indoor to outdoor cultivation systems. Indoor cultures include controlled laboratory methods that allow control of temperature, pH, illumination, and nutrients, whereas in outdoor cultures it is difficult to grow microalgae without the risk of contamination. The most routinely used systems are continuous, semicontinuous, and batch cultures, as well as open or closed ponds and photobioreactors.

9.4 WHAT ARE MICROALGAL CULTIVATION TECHNIQUES FOR ENVIRONMENTAL CLEANUP?

One of the most significant anthropogenic and natural problems throughout history is air pollution. The main components of flue gas are carbon oxides (CO_x), sulfur oxides (SO_x), nitrogen oxides (NO_x), hydrocarbons, trace amounts of heavy metals, and particulate matter, all of which have harmful effects on the ecosystem.

CO_2 levels exceeding 400 ppm cause climate change (Duarte et al., 2012). On the other hand, the effects of car exhaust emissions on air pollution represent another major problem the world is facing today. There is serious concern about traffic-related air pollution in urban areas due to its contribution to greenhouse gas levels. CO_2 emissions are among the most important greenhouse gas–related environmental problems. The products of incomplete combustion are CO and hydrocarbons, which are additional environmental concerns. Carbon dioxide emissions and gasoline consumption per 100 km of driving have been calculated as 21.1 kg CO_2 and 5.47 kg gasoline, respectively (Matsumoto et al., 1997a).

Microalgae have the potential to provide solutions to many of the environmental problems we face today. They have been widely and successfully utilized in air pollution studies due to their high CO_2 utilization ability (Jacob-Lopes et al., 2010; Murray et al., 2012; Radmann et al., 2011). The use of microalgae in various industrial facilities and their products is summarized in Figure 9.3, particularly in terms of CO_2 cleanup technology and biomass production. Microalgae can fix carbon from different sources, including (Wang et al., 2008): (i) the atmosphere; (ii) industrial exhaust gases; and (iii) soluble carbonate forms such as $NaHCO_3$ and Na_2CO_3.

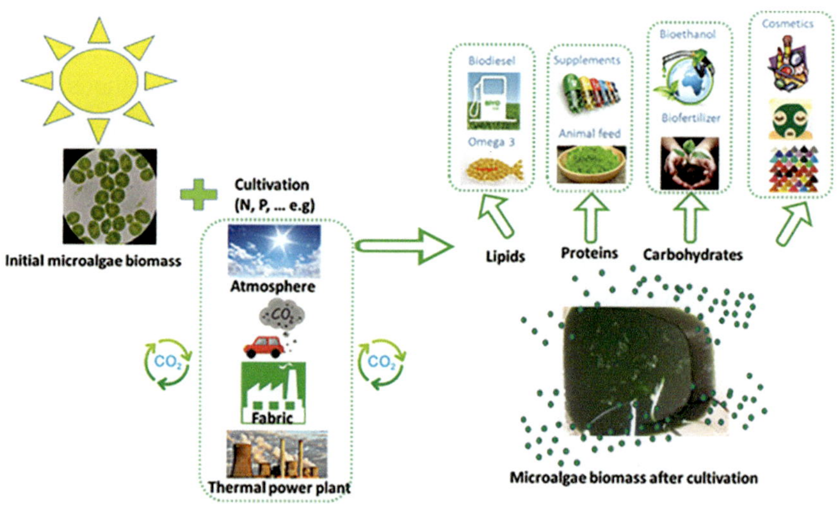

FIGURE 9.3 Schematic diagram of microalgal cultivation as CO_2 cleanup technology and biomass production.

The atmosphere contains many flue gas components (SO_x, NO_x, and CO_x), metallic particles (lead, cadmium, nickel, asbestos), small and large particulate matter, ozone and other photochemical oxidizers, hydrocarbons, and polycyclic aromatic hydrocarbons (PAHs). Flue gas components (SO_x, NO_x, and CO_x) and their effects on microalgae have been detailed above.

9.4.1 SO_x

SO_x is a colorless, pungent-smelling gas. It has high water solubility, 22.971 g/100 g H_2O at 0°C and 5.881 g/100 g H_2O at 40°C (Weil and Sandler, 1997). It oxidizes in the presence of water droplets in the air. In general, the distribution of SO_x and NO_x ratios in flue gas is 98% SO_2 and 90%–95% NO (Dahlan et al., 2011; Pillai et al., 2009). The reactions of SO_2 released from flue gases into the air are as follows:

$$S + O_2 \rightarrow SO_2$$

SO_2 can form H_2SO_3 after dissolving in water (Stumm and Morgan, 1981):

$$SO_2 + H_2O \rightarrow H_2SO_3$$

In this process, the main oxidized form is SO_3. SO_3 is present in flue gas at 2%–4% (Niessen, 2002), and when it reacts with water, it quickly forms sulfuric acid (H_2SO_4). This transformation is shown below. H_2SO_4 is a very strong acid.

$$SO_3 + H_2O \rightarrow H_2SO_4$$

Another sulfur compound that causes environmental pollution is hydrogen sulfide (H_2S). Ammonium bisulfate (NH_4HSO_4) is also a sulfur-containing compound that is less acidic than sulfuric acid. As a result of the reaction of SO_x and NO_x derivatives with water vapor in the atmosphere, acid aerosols or acid rain are formed. The presence of SO_2 is primarily due to industrial activities and is one of the most important contributors to air pollution. In addition, some SO_2 is released into the air as a result of forest fires and volcanic eruptions. The majority of the sulfur released into the atmosphere results from the burning of fossil fuels. The reduction of sulfur content in fuels used today through various processes has significantly reduced the SO_2 concentration in such fuels. Every hour, 0.5%–10% of SO_2 emissions resulting from combustion are converted into sulfur trioxide (SO_3). Sulfur dioxide can naturally be found in air at concentrations of less than 5 μg/m³. The amount of H_2SO_4 in indoor air is negligibly low. In urban areas in Europe, the average annual sulfur dioxide concentration is generally around 100–200 μg/m³. The daily average concentration varies between 250 μg/m³ and 500 μg/m³. One-hour maximum measurement values can range between 1000 μg/m³ and 2000 μg/m³ (RSHC, 2004). Sulfur dioxide affects a large population worldwide. Lung damage caused by SO_2 occurs at concentrations of <0.5 ppm, and this acidic gas also has an irritant effect (van Thriel et al., 2010). About 85% of sulfur dioxide enters the body through the airways and enters the bloodstream. It is converted into sulfate in the liver and is excreted from the body.

Sulfur compounds exert their primary and most significant effects on the respiratory system. As a result, lung diseases such as bronchitis may occur. Acid aerosols and particulate matter are transported from the lungs to the alveoli; therefore, the combined negative health effects of these pollutants are greater than the effects of each individually. Major health problems arising from these effects include pulmonary dysfunction, increased cases of chronic bronchitis, impaired clearance of bronchial mucosal cilia, and thickening of the respiratory epithelial tissue (EPA, 2020a). The effects of sulfur dioxide on microalgae have been reported by Darall (1989).

As expected, the effects are dose-dependent and have been reported to show no significant inhibitory effects below 500 ppb for exposure durations of less than eight hours. Reported damages include decreases in photosynthetic pigments, enzyme activities, food intake, and growth; decreased productivity; physiological inhibition; changes in metabolic functions; and damage to chloroplasts, pigments, and thylakoid membranes due to photooxidation. On the other hand, increased CO_2 concentrations have been found to protect against the inhibitory effects of SO_2 on photosynthesis, which has been linked by researchers to partial closure of stomata.

9.4.2 MICROALGAL SO_X TOLERANCES FROM FLUE GAS

Sulfur (S) is a major component of the amino acids cysteine and methionine and is incorporated into the structure of sulfur-containing thylakoid lipids, which are essential for microalgal development (Graham and Wilcox, 2000). Freshwater algae contain 0.15%–1.96% w/w S per dry cell weight. In this context, microalgae require S for sulfate formation in their cytoplasm (Giordano et al., 2005a, b). Sulfate is transported to plastids, and excess sulfate is stored in vacuoles. This process involves activation to 5′-adenylyl sulfate (APS). The produced sulfite (SO_3^{2-}) is reduced to sulfide (S^{2-}) by sulfite reductase and is immediately incorporated into the structure of cysteine (Van Den Hende et al., 2012), as shown in Figure 9.4. *Emiliania* sp. and *Phaeocystis* sp. use sulfur compounds as a cold-protection mechanism (Graham and Wilcox, 2000).

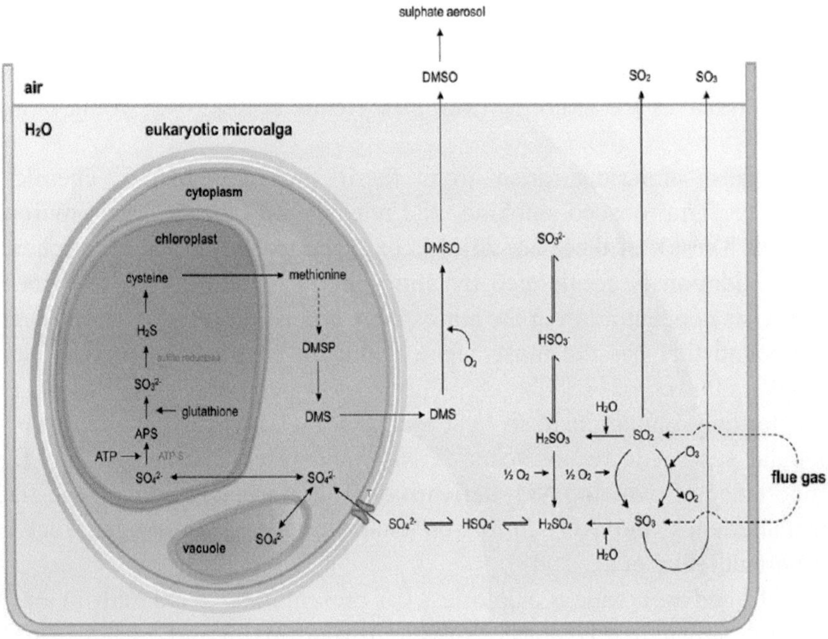

FIGURE 9.4 Sulfur uptake and storage pathways of microalgae and cyanobacteria.
Source: Adapted from Van Den Hende et al. (2012).

9.4.3 NO_X

Nitrogen oxides, another group of air pollutants, are divided into nitrogen dioxide (NO_2) and nitric oxide (NO). Other NO_x derivatives present in flue gas at very low concentrations include N_2O_3 (black solid), N_2O_4 (red-brown gas), N_2O_2 (colorless gas), and N_2O_5 (white solid) (Koppejan and van Loo, 2007). If these NO_x derivatives are water-soluble, they can form nitric acid (HNO_3) or nitrous acid (HNO_2) (Van Den Hende et al., 2012). In general, the most common NO_x components in flue gas are 95% NO and

5% NO_2 (Wang et al., 2008). Nitric acid is a highly harmful acid because it is both a strong acid and a strong oxidant. In combustion, NO_x can occur in four different ways (Van Den Hende et al., 2012):

- Thermal NO_x; occurs at very high temperatures;
- Sudden NO_x; occurs during sudden temperature changes involving nitrogen in the atmosphere;
- Fuel NO_x; consists of nitrogen present in the fuel;
- NO_x from N_2O.

Only 5% of the NO_x generally formed is oxidized to NO_2 before the emission leaves the chimney (Van Den Hende et al., 2012; Wang et al., 2008).

Exhaust emissions arise from fossil fuel combustion, chemical processes, and tobacco smoking. NO released into the external environment as a result of these activities is oxidized to NO_2 in the atmosphere. This oxidation is accelerated by atmospheric oxidizers such as ozone. Due to its concentration in the atmosphere and its properties, the nitrogen compound that has the most negative impact on human health is NO_2 (RSHC, 2004).

The only contaminant in hydrogen (H) fuels, which have become a popular subject in recent years, is NO_x, and researchers emphasize the importance of reducing NO_x derivatives. Therefore, determining the fuel characteristics and NO_x formation characteristics in hydrogen fuel is important (Ilbas et al., 2005).

Nitrogen monoxide is a colorless gas present in trace amounts in air. It is a free radical and is slightly soluble in water (0.032 g/L, 1 atm, 25°C; Dora et al., 2009). NO_2 is a reddish-brown gas formed by the oxidation of NO:

$$2NO \rightarrow N_2 + O_2$$

$$N_2O_2 + O_2 \rightarrow O_2 + 2NO_2$$

NO_2 has 6000 times higher solubility than NO (213.0 g/L, 1 atm, 25°C; Dora et al., 2009).

The concentration of NO_2 in urban areas varies according to the hours of the day, seasons, and meteorological conditions. In cities with heavy traffic, a high NO_2 concentration such as 0.5 ppm can be measured as an hourly average value. As a result of smoking and the use of liquid and gas fuels for heating and cooking, individual NO_2 exposure increases. As a result of the use of unvented heating appliances, the NO_2 concentration in indoor environments can reach much higher levels than in outdoor environments (RSHC, 2004). NO_x plays a role in the nitrogen cycle. In addition, through biological nitrogen fixation, nitrogen is converted into nitrates or ammonia that can be used by living organisms. This transformation occurs via nitrifying bacteria.

9.4.4 EFFECTS OF NO_X ON MICROALGAE

Photosynthesis is less sensitive to the presence of NO than to SO_2 (Darall, 1989). No effect was observed for exposure durations of less than eight hours at concentrations not exceeding 500–700 ppb. NO is reported to be four times more inhibitory than NO_2. The combined effect of $SO_2 + NO_x$ leads to greater inhibition than the individual effects.

9.4.5 FLUE GAS MICROALGAL NO_X TOLERANCE

Microalgal NO_x tolerance depends on the initial microalgal cell concentration (Yoshihara et al., 1996), NO_x concentration (Doucha et al., 2005), NO_x gas flow rate (Li et al., 2011), and species (Radmann and Costa, 2008). As reported, the effect of NO_x on microalgae has not yet been fully clarified (Van Den Hende et al., 2012).

Jin et al. (2008) added chelators to the medium to enhance NO_x removal in microalgal cultures. However, Fe(II)-EDTA, one of the best-known chelates, is oxidized by dissolved O_2 in flue gas and microalgal cultures and transformed into Fe(III)-EDTA, thereby decreasing its performance. This process is very expensive, and EDTA micropollutants may harm cultures (Van Den Hende et al., 2012). Figure 9.5 shows the mechanism of microalgal NO_x tolerance in eukaryotic and prokaryotic cells (Van Den Hende et al., 2012).

9.4.6 CARBON OXIDES (CO_X)

In terms of air pollution, two important carbon compounds are carbon monoxide (CO) and carbon dioxide (CO_2).

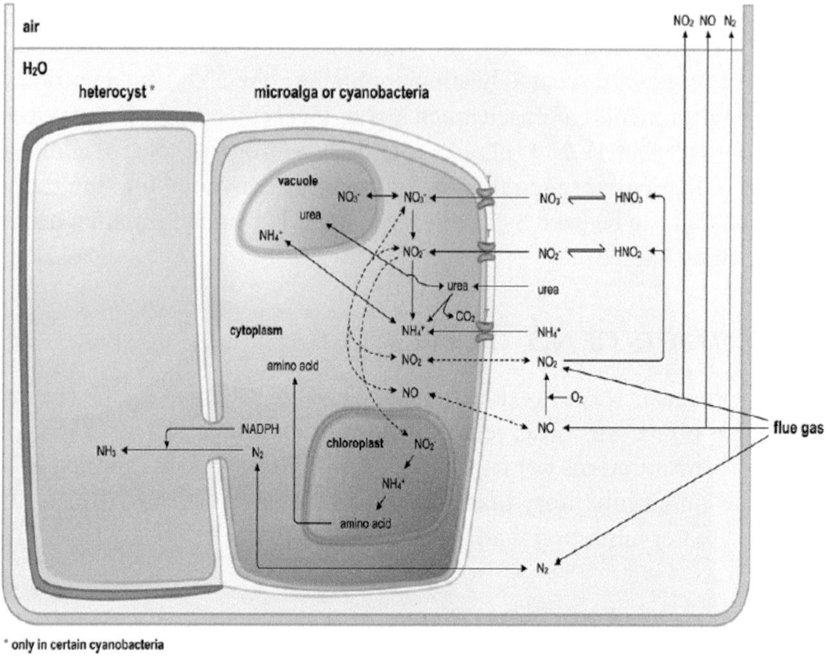

FIGURE 9.5 Nitrogen uptake and storage pathways of microalgae and cyanobacteria.
Source: Adapted from Van Den Hende et al. (2012).

9.4.7 CARBON MONOXIDE (CO)

Carbon monoxide is an air pollutant commonly found in the atmosphere as a result of incomplete combustion. It is a colorless, odorless, water-soluble gas that is lighter than air and highly toxic. Due to the increased use of vehicles, its concentration in the atmosphere increases steadily. Fossil fuels, vehicles, and industrial and biological processes are the most important sources of carbon monoxide. Carbon monoxide from natural sources is negligible. Flue gas contains CO and unburned hydrocarbons (C_xH_y), such as methane (CH_4) and butane (C_4H_{10}), formed during incomplete combustion due to insufficient air supply or low temperature (Trozzi et al., 2010). Carbon monoxide remains in the atmosphere longer than many other pollutants. The conversion reactions of carbon monoxide to carbon dioxide in the atmosphere are as follows:

$$CO + O_2 \rightarrow CO_2 + O$$

$$CO + O_3 \rightarrow CO_2 + O_2$$

$$CO + H_2O \rightarrow CO_2 + H_2$$

The natural level of CO is 0.001–0.20 ppm. Its concentration in urban areas varies according to weather conditions, traffic density, time of day, and distance from its sources. The average maximum 8-hour concentration in urban areas has been reported to be 53 ppm. Peak concentrations are observed in the morning and evening when traffic is heavy. Particularly high CO concentrations are detected inside moving vehicles, in garages, and in areas with heavy traffic flow. Therefore, high CO exposure is especially expected among drivers, garage workers, and traffic police. For example, measurements have shown maximum concentrations of 36 ppm inside vehicles and 500 ppm in garages. The use of stoves for indoor heating and the burning of fossil fuels under improper conditions significantly increase CO concentrations. Cooking processes are also important indoor sources. Measurements in these environments have shown that CO concentrations vary between 10 ppm and 50 ppm. Tobacco smoking is one of the most important indoor sources of CO (RSHC, 2004).

Dermal exposure to CO is not significant; the primary route of exposure is through the respiratory tract. If its concentration in the atmosphere increases, CO passes into the bloodstream and binds to hemoglobin. Its affinity for hemoglobin is 200 times greater than that of oxygen. This is the main reason for stove poisoning. When oxygen cannot bind to the iron in hemoglobin, oxyhemoglobin cannot form, resulting in the formation of carboxyhemoglobin in the blood. Health problems associated with CO exposure are listed below (RSHC, 2004):

- Cardiovascular effects;
- Neurological effects;
- Fibrinolytic effects;
- Perinatal effects.

As a result of the decrease in oxygen-carrying capacity due to CO exposure, dysfunction occurs in sensitive organs and tissues such as blood vessel walls, the brain, and the heart because of oxygen deficiency in the blood. The amount of carboxyhemoglobin (COHb) in the blood recommended by the World Health Organization is around 2.5%–3.0% (WHO, 2000).

9.4.8 CARBON DIOXIDE (CO_2)

Carbon dioxide (CO_2) is known as the most important greenhouse gas, and its concentration in the atmosphere has increased due to rising industrial activities and vehicle use, especially in the last 25 years (EPA, 2020b). It is a colorless and odorless gas. One million air molecules contain about 350 CO_2 molecules. It plays a key role in respiration and photosynthesis.

Increasing concentrations of CO_2 in the atmosphere act like a glass enclosure. Short-wavelength solar radiation passes easily through the atmosphere, but the increased heat energy inside is trapped by CO_2 and cannot be released back into space, causing the internal temperature to rise. In this context, the greenhouse effect occurs when radiation reflected from the earth is retained by atmospheric gases, especially CO_2, methane, and water vapor, thereby increasing the earth's average temperature. Although carbon resources on earth remained stable for thousands of years, modern human activities—such as the use of fossil fuels, deforestation, and excessive agricultural practices—have increased the amount of CO_2 and other greenhouse gases in the atmosphere, thereby intensifying the greenhouse effect. Global warming is the increase in the earth's average atmospheric temperature resulting from the greenhouse effect. The atmospheric CO_2 concentration, previously about 280 ppm, has increased to approximately 350 ppm due to the increasing use of fossil fuels (Flexas et al., 2012).

The amount of CO_2 in flue gas varies according to the type of fuel. For example, in coal-fired power plants, the CO_2 concentration is around 10%–15%, whereas it is around 5%–6% for natural gas–fired systems (Van Den Hende et al., 2012). The functioning of microalgal CO_2 tolerance in eukaryotic and prokaryotic cells is shown in Figure 9.6 (Van Den Hende et al., 2012).

9.4.9 MICROALGAL CARBON DIOXIDE (CO_2) TOLERANCE FROM FLUE GAS

CO_2 easily diffuses across the plasma membrane of microalgae (Amoroso et al., 1998). Ribulose-1,5-bisphosphate carboxylase/oxygenase (Rubisco) is the first enzyme of the Calvin cycle and converts assimilated CO_2 into 3-phosphoglycerate (3-PGA). In addition to its carboxylase activity, Rubisco also exhibits oxygenase activity and catalyzes the reaction of O_2, leading to the formation of 3-PGA and phosphoglycolate. Due to the

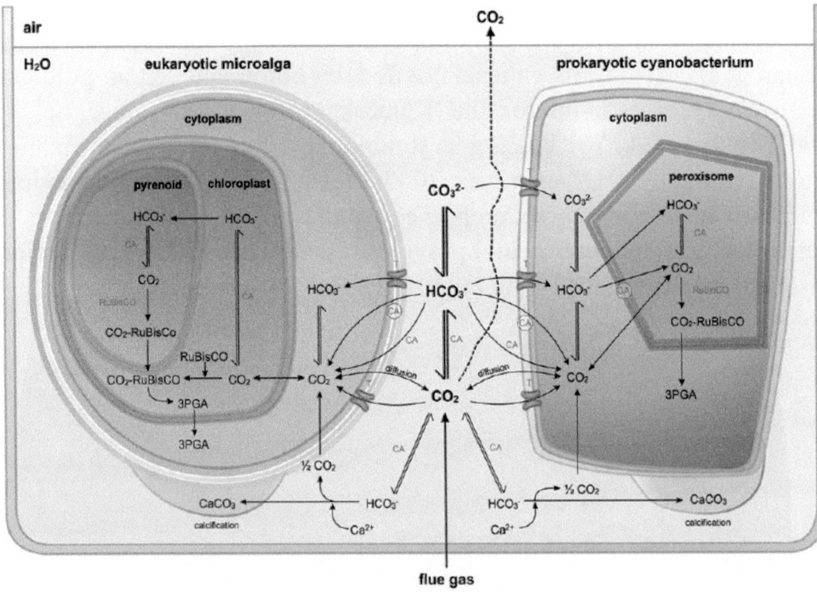

FIGURE 9.6 Inorganic C uptake and storage pathways of microalgae and cyanobacteria. *Source:* Adapted from Van Den Hende et al. (2012).

low carbon-binding capacity of Rubisco, microalgae have developed a carbon-concentrating mechanism (CCMs). Through this mechanism, the intracellular CO_2 concentration can increase up to 1000 times relative to the surrounding environment (Price et al., 2008).

Many types of microalgae can tolerate CO_2 concentrations of 5%–20% in flue gas (Taştan et al., 2013, 2016a). In addition, microalgae capable of tolerating CO_2 concentrations between 40% and 100% have been reported in the literature (Matsumoto et al., 1997b; Olaizola et al., 2003). Many factors, such as low emission flow rates, the CCM mechanism, the addition of a base to the medium, and the prevention of culture acidification, contribute to this high tolerance capacity.

9.4.10 RESISTANCE FACTORS

9.4.10.1 TRIACONTANOL

1-Triacontanol is a fatty alcohol with the general formula $C_{30}H_{62}O$, also known as melissyl alcohol or myricyl alcohol (Figure 9.7). It is a wax-like plant hormone found in the cuticle layer of plants. The stimulating effect of triacontanol on growth and chlorophyll concentration (Kiliç et al., 2010; Taştan et al., 2012a), its enhancing effect on boron and certain pollutant removal capacities in microalgae (Karacakaya et al., 2009; Naeem et al., 2009; Taştan et al., 2012a), and its stimulating effect on nitrogen fixation and enzyme activity (Naeem et al., 2009) have been highlighted in previous studies. Considering these properties, studies have been conducted to determine whether triacontanol provides resistance to microalgae against toxic gases in flue gas (Taştan et al., 2012b).

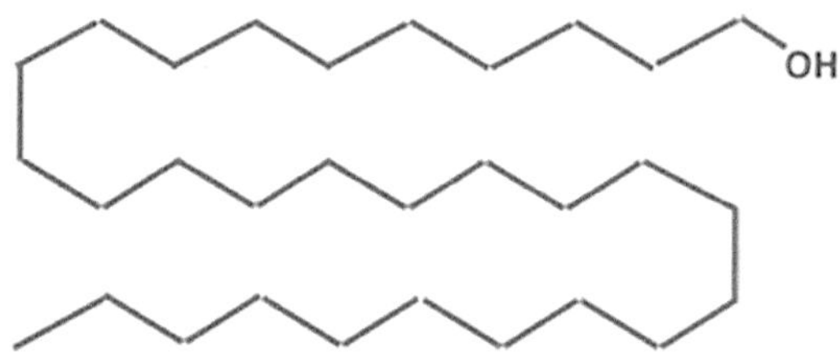

FIGURE 9.7 Chemical structure of 1-triacontanol.

9.4.10.2 SALICYLIC ACID

Salicylic acid is a phenolic acid named after *Salix alba* (willow tree), from whose bark it was first derived. Chemically, it is a monohydroxybenzoic acid and a beta-hydroxy acid (Figure 9.8). It is a plant hormone used in microalgal flue gas capture studies (Taştan et al., 2013, 2016a).

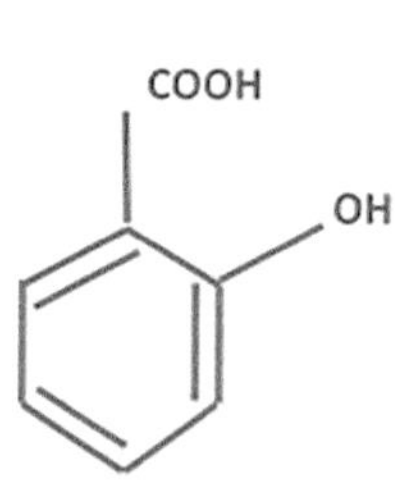

FIGURE 9.8 Chemical structure of salicylic acid.

Salicylic acid was reportedly obtained by Hippocrates about 3500 years ago from the willow tree. It is the precursor of aspirin (acetylsalicylic acid) in its current pharmaceutical use and is also widely known as an anti-acne treatment agent. Salicylic acid is derived from salicin metabolism. Salicylic acid salts and esters are known as "salicylates."

Salicylic acid is a well-known example of phenolic acids. In addition to being an important signaling molecule, it also has properties such as resistance enhancement and growth stimulation (Hayat et al., 2010). Although the analgesic effect of salicylic acid obtained from willow bark has been known for many years, aspirin was synthesized 160 years ago (Lafont, 2007). Although studies on the effects of salicylic acid in plants are limited, available data emphasize that this substance has an important role among plant growth hormones, particularly in signaling mechanisms and tolerance to biotic and abiotic stress factors (Rivas et al., 2011).

9.4.10.3 SODIUM BICARBONATE ($NAHCO_3$)

$NaHCO_3$, or sodium hydrogen carbonate, is a crystalline chemical compound (Figure 9.9). The natural mineral forms are called nahcolite and natron. Sodium bicarbonate is known to enhance resistance in microalgal cells and assist them in tolerating flue gas (Taştan et al., 2012b). Microalgae can absorb CO_2 from three different sources (Wang et al., 2008):

- From the atmosphere;
- Industrial exhaust or flue gases;
- Soluble carbonate forms ($NaHCO_3$, Na_2CO_3).

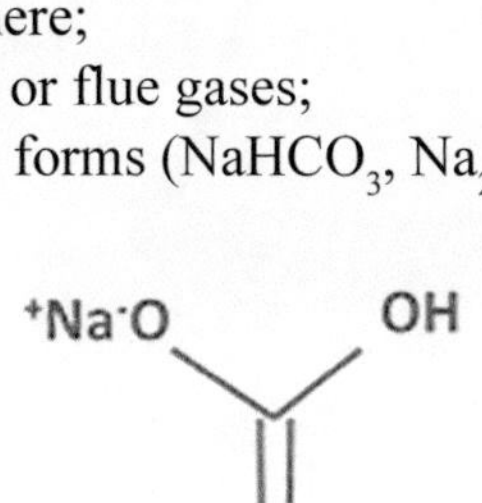

FIGURE 9.9 Chemical structure of sodium bicarbonate ($NaHCO_3$).

9.4.11 FLOW RATE

Flow rate is expressed as the amount of flue gas used per minute. The effects of gas flow rates from coal emissions, LPG, and gasoline exhaust emissions on microalgal growth have been investigated by many researchers (Taştan et al., 2013, 2016a) within the range of 0.6–2.4 vvm. Microalgal cultures are aerated daily with these emissions at known flow rates from the 5th to the 15th day of incubation. An increase in flow rate results in greater microalgal biomass production (Taştan et al., 2013). In some studies, researchers produced artificial flue gas using their own setup (Figure 9.10) and supplied it to microalgal cultures (Taştan et al., 2012b).

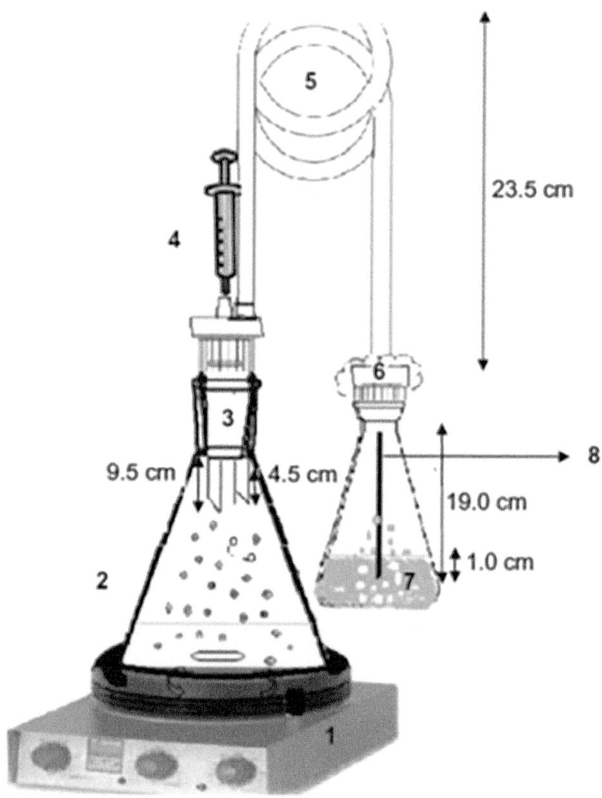

FIGURE 9.10 Artificial flue gas transfer system in microalgae.
Note: (1) Magnetic heat stirrer; (2) Erlenmeyer flask; (3) gum elastic dowel with acid inlet and gas exit ports; (4) syringe; (5) heat-resistant hose; (6) gum elastic dowel with cotton; (7) Erlenmeyer flask with microalgal culture (250 mL); and (8) needle.

Source: Adapted from Taştan et al. (2012b).

9.5 APPLICATIONS OF MICROALGAL CLEANUP TECHNOLOGY AND FUTURE PERSPECTIVES

There are many areas where microalgae are used in renewable energy studies. One of them is C-capturing studies. The atmospheric CO_2 concentration of about 400 ppmv is not adequate for large-scale microalgal production; the actual requirement is 1.3–2.4 kg CO_2 to obtain 1 kg of dry weight of microalgae (Becker, 2008). Therefore, using C-capturing techniques in flue gas emission studies with microalgae is an effective way to produce microalgal biomass. In addition, the CO_2 and NO_x mixture is an effective C and N source for microalgal growth (Kao et al., 2014).

When it comes to renewable energy, microalgae-based bioenergy studies come first. In some studies, CO_2-capturing techniques in coal emissions by bio-stimulated *Chlorella* sp. were well demonstrated through the enhancement of its growth and resistance using exogenous salicylic acid and 1-triacontanol growth promoters (Taştan et al., 2012b, 2013). Taştan et al. (2012b) and Wang et al. (2008) found that $NaHCO_3$ was a suitable C source for the growth of *Chlorella vulgaris*, *Chlorella* sp., and *Synechococcus* sp. In addition, pH is a very important parameter in the cultivation of microalgae. Radmann et al. (2011) stated that pH is closely related to inorganic carbon availability in the medium during the biofixation of CO_2. Besides this, de Morais and Costa (2007) revealed that biomass increase is directly related to the pH value during CO_2 biofixation by *C. kessleri*.

Flue gases contain several chemicals such as O_2, CO_2, NO_x, SO_x, CO, HCs, heavy metals, and other compounds, which can affect the growth of microalgae. Some types of microalgae may be more resistant than others in flue gas treatment and CO_2-capturing studies. Several microalgal genera have been studied for CO_2 capture, such as *Chlorella* (Arbib et al., 2014), *Scenedesmus* (Kim et al., 2013), and *Spirulina* species (Zeng et al., 2012). According to the findings of Radmann et al. (2011) and Taştan et al. (2012b), *Chlorella* sp. is more resistant to toxic gases in flue gas than *Synechococcus* sp. Similarly, Douskova et al. (2009) clarified that higher CO_2 levels (11% v/v) did not show any inhibitory effects on *C. vulgaris*. *Chlorella* sp. can be regarded as one of the most flue-gas-tolerant microalgal species.

Even in some studies on the utilization of flue gas and CO_2 capture, microalgal technologies are considered the most promising methods for CO_2 capture and utilization (Arbib et al., 2014; Taştan et al., 2012b, 2013, 2016a, b).

9.6 CONCLUSION

This chapter reports the concepts of algal cultivation techniques for current and future environmental cleanup. In this chapter, the advantages and challenges of algal cultivation techniques by means of a carbon-capture strategy are discussed. CO_2 is the main contributor to the global warming problem. Among various C-capture techniques, microalgae-based C capture is one of the promising clean energy technologies. C capture is one of the dominant strategies to reduce climate change and greenhouse effects. Most microalgal species can be used as effective biomaterials for flue gas utilization, and their performance can be enhanced by adding plant growth stimulation factors to the culture media. CO_2 can be utilized by microalgae and converted into biomass through an efficient cultivation method. Capturing C by microalgae from the environment can serve as an environmentally sustainable process.

KEYWORDS

- **air pollution**
- **carbon dioxide emissions**
- **cyanobacteria**
- **flue gas**
- **green algae**
- **greenhouse gases**
- **microalgae**
- **microalgal cultivation**

REFERENCES

Amoroso, G., Sültemeyer, D., Thyssen, C., & Fock, H. P. (1998). Uptake of HCO_3 and CO_2 in cells and chloroplasts from the microalgae *Chlamydomonas reinhardtii* and *Dunaliella tertiolecta*. *Plant Physiology, 116*, 193–201.

Arbib, Z., Ruiz, J., Àlvarez-Díaz, P., Garrido-Pèrez, C., & Perales, J. A. (2014). Capability of different microalgae species for phytoremediation processes: Wastewater tertiary treatment, CO_2 bio-fixation and low cost biofuels production. *Water Research, 49*, 465–474.

Barsanti, L., & Gualtieri, P. (2014). *Algae: Anatomy, Biochemistry, and Biotechnology*. Boca Raton, FL: CRC Press Taylor and Francis Group.

Becker, E. W. (2008). *Microalgae: Biotechnology and Microbiology*. Cambridge: Cambridge University Press.

Chae, S. R., Hwang, E. J., & Shin, H. S. (2006). Single cell protein production of *Euglena gracilis* and carbon dioxide fixation in an innovative photobioreactor. *Bioresource Technology, 97*, 322–329.

Dahlan, I., Lee, K. T., Kamaruddin, A. H., & Mohamed, A. R. (2011). Sorption of SO_2 and NO from simulated flue gas over rice husk ash (RHA)/CaO/CeO_2 sorbent: Evaluation of deactivation kinetic parameters. *Journal of Hazardous Materials, 185*, 1609–1613.

Darall, N. M. (1989). The effect of air pollutants on physiological processes in plants. *Plant, Cell and Environment, 12*, 1–30.

Dayananda, C., Sarada, R., Usha Rani, M., Shamala, T. R., & Ravishankar, G. A. (2007). Autotrophic cultivation of *Botryococcus braunii* for the production of hydrocarbons and exopolysaccharides in various media. *Biomass and Bioenergy, 31*(1), 87–93.

de Morais, M. G., & Costa, J. A. V. (2007). Isolation and selection of microalgae from coal fired thermoelectric power plant for biofixation of carbon dioxide. *Energy Conversion and Management, 48*, 2169–2173.

de-Bashan, L. E., & Bashan, Y. (2010). Immobilized microalgae for removing pollutants: Review of practical aspects. *Bioresource Technology, 101*, 1611–1627.

Dora, J., Gostomczyk, M. A., Jakubiak, M., Kortylewski, W., Mista, W., & Tkaczuk, M. (2009). Parametric studies of the effectiveness of oxidation of NO by ozone. *Chemical and Process Engineering, 30*, 621–634.

Doucha, J., Straka, F., & Livansky, K. (2005). Utilization of flue gas for cultivation of microalgae (*Chlorella* sp.) in an outdoor open thin-layer photobioreactor. *Journal of Applied Phycology, 17*, 403–412.

Douskova, I., Doucha, J., Livansky, K., et al. (2009). Simultaneous flue gas bioremediation and reduction of microalgal biomass production costs. *Applied Microbiology and Biotechnology, 82*, 179–185.

Duarte, C. M., Lenton, T. M., Wadhams, P., & Wassmann, P. (2012). Abrupt climate change in the Arctic. *Nature Climate Change, 2*, 60–62.

EPA. (2020a). https://www.epa.gov/so2-pollution (accessed on 26th February 2026).

EPA. (2020b). https://www.epa.gov/ghgemissions/overview-greenhouse-gases (accessed on 26th February 2026).

Flexas, J., Loreto, F., & Medrano, H. (2012). *Terrestrial Photosynthesis in a Changing Environment: A Molecular, Physiological, and Ecological Approach.* Cambridge: Cambridge University Press.

Giordano, M., Beardall, J., & Raven, J. A. (2005b). CO_2 concentrating mechanisms in algae: Mechanisms, environmental modulation, and evolution. *Annual Review of Plant Biology, 56*, 99–131.

Giordano, M., Norici, A., & Hell, R. (2005a). Sulfur and phytoplankton: Acquisition, metabolism and impact on the environment. *The New Phytologist, 166*, 371–382.

Graham, L. E., & Wilcox, L. W. (2000). *Algae*. Upper Saddle River, New Jersey: Prentice Hall.

Hayat, Q., Hayat, S., Irfan, M., & Ahmad, A. (2010). Effect of exogenous salicylic acid under changing environment: A review. *Environmental and Experimental Botany, 68*, 14–25.

IEA (International Energy Agency). (2012). *World Energy Outlook*. https://iea.blob.core.windows.net/assets/ebe15dfb-30c8-42cf-8733-672b3500aed7/WEO2012_free.pdf (accessed on 26th February 2026).

Ilbas, M., Yilmaz, I., & Kaplan, Y. (2005). Investigations of hydrogen and hydrogen–hydrocarbon composite fuel combustion and image emission characteristics in a model combustor. *International Journal of Hydrogen Energy, 30*, 1139–1147.

Jacob-Lopes, E., Scoparo, C. H. G., Queiroz, M. I., & Franco, T. T. (2010). Biotransformations of carbon dioxide in photobioreactors. *Energy Conversion and Management, 51*, 894–900.

Jin, H. F., Santiago, D. E. O., Park, J., & Lee, K. (2008). Enhancement of nitric oxide solubility using Fe(II) EDTA and its removal by green algae *Scenedesmus* sp. *Biotechnology and Bioprocess Engineering, 13*, 48–52.

Kao, C. Y., Chen, T. Y., Chang, Y. B., et al. (2014). Utilization of carbon dioxide in industrial flue gases for the cultivation of microalga *Chlorella* sp. *Bioresource Technology, 166*, 485–493.

Karacakaya, P., Kılıç, N. K., Duygu, E., & Dönmez, G. (2009). Stimulation of reactive dye removal by cyanobacteria in media containing triacontanol hormone. *Journal of Hazardous Materials, 172*, 1635–1639.

Kiliç, N. K., Duygu, E., & Dönmez, G. (2010). Triacontanol hormone stimulates population, growth and Brilliant Blue R dye removal by common duckweed from culture media. *Journal of Hazardous Materials, 182*, 525–530.

Kim, G., Choi, W., Lee, C. H., & Lee, K. (2013). Enhancement of dissolved inorganic carbon and carbon fixation by green alga *Scenedesmus* sp. in the presence of alkanolamine CO_2 absorbents. *Biochemical Engineering Journal, 78*, 18–23.

Koppejan, J., & van Loo, S. (2007). *The Handbook of Biomass Combustion and Co-Firing*. London: Routledge.

Lafont, O. (2007). From the willow to aspirin. *Revue d'histoire de la Pharmacie, 55*, 209–216.

Li, F. F., Yang, Z. H., Zeng, R., et al. (2011). Microalgae capture of CO_2 from actual flue gas discharged from a combustion chamber. *Industrial and Engineering Chemistry Research, 50*, 6496–6502.

Machida, T., Kita, K., Kondo, Y., et al. (2003). Vertical and meridional distributions of the atmospheric CO_2 mixing ratio between northern midlatitudes and southern subtropics. *Journal of Geophysical Research, 108*, 1–9.

Matsumoto, H., Hamasaki, A., Sioji, N., & Ikuta, Y. (1997b). Influence of CO_2, SO_2 and NO in flue gas on microalgae productivity. *Journal of Chemical Engineering of Japan, 30*, 620–624.

Matsumoto, S., Inaba, A., & Yanagisawa, Y. (1997a). Technology assessment of alternative fuels by CO_2 fixation use in passenger cars. *Energy Conversion and Management, 38*, 455–460.

Murray, K. E., Shields, J. A., Garcia, N. D., & Healy, F. G. (2012). Productivity, carbon utilization, and energy content of mass in scalable microalgae systems. *Bioresource Technology, 114*, 499–506.

Naeem, M., Khan, M. M. A., Moinuddin, & Siddiqui, M. H. (2009). Triacontanol stimulates nitrogen-fixation, enzyme activities, photosynthesis, crop productivity and quality of hyacinth bean (*Lablab purpureus* L.). *Scientia Horticulture, 121*, 389–396.

Niessen, W. R. (2002). *Combustion and Incineration Processes*. New York: Marcel Dekker Inc.

Olaizola, M., Bridges, T., Flores, S., Griswold, L., Morency, J., & Nakamura, T. (2003). Microalgal removal of CO_2 from flue gases: Changes in medium pH and flue gas composition do not appear to affect the photochemical yield of microalgal cultures. *Biotechnology and Bioprocess Engineering, 8*, 360–367.

Pillai, K. C., Chung, S. J., & Raju, T. (2009). Experimental aspects of combined NOx and SO_2 removal from flue-gas mixture in an integrated wet scrubber–electrochemical cell system. *Chemosphere, 76*, 657–664.

Price, G. D., Badger, M. R., Woodger, F. J., & Long, B. M. (2008). Advances in understanding the cyanobacterial CO_2-concentrating mechanisms (CCM): Functional components, Ci transporters, diversity, genetic regulation, prospects for engineering into plants. *Journal of Experimental Botany, 59*, 1444–1461.

Radmann, E. M., & Costa, J. A. V. (2008). Conteúdo lipídico e composição de ácidos graxos de microalgas expostas aos gases CO_2, SO_2 e NO. *Química Nova, 31*, 1609–1612.

Radmann, E. M., Camerini, F. V., Santos, T. D., & Costa, J. A. V. (2011). Isolation and application of SOx and NOx resistant microalgae in biofixation of CO_2 from thermoelectricity plants. *Energy Conversion and Management, 52*, 3132–3136.

Rippka, R. (1988). Recognition and identification of cyanobacteria. *Methods in Enzymology, 167*, 28–67.

Rivas, M., Vicente, S., & Plasencia, J. (2011). Salicylic acid beyond defense: Its role in plant growth and development. *Journal of Experimental Botany, 62*, 3321–3338.

RSHC (Environmental Health Research Directorate of Refik Saydam Hygiene Center). (2004). An overview of air pollution Ankara, Türkiye. https://www.yumpu.com/tr/document/read/15838878/hava-kirliligine-genel-baks-refik-saydam-hfzsshha-merkezi-pdf (accessed on 26th February 2026).

Stumm, W., & Morgan, J. J. (1981). *Aquatic Chemistry*. New York: John Wiley and Sons.

Sydney, E. B., Sturm, W. S., de Carvalho, J. C., et al. (2010). Potential carbon dioxide fixation by industrially important microalgae. *Bioresource Technology, 101*, 5892–5896.

Taştan, B. E., & Tekinay, T. (2016b). A novel coal additive from microalgae produced from thermal power plant flue gas. *Journal of Cleaner Production, 133*, 1086–1094.

Taştan, B. E., Duygu, E., & Dönmez, G. (2012a). Boron bioremoval by a newly isolated *Chlorella* sp. and its stimulation by growth stimulators. *Water Research, 46*, 167–175.

Taştan, B. E., Duygu, E., Atakol, O., & Dönmez, G. (2012b). SO_2 and NO_2 tolerance of microalgae with the help of some growth stimulators. *Energy Conversion and Management, 64*, 28–34.

Taştan, B. E., Duygu, E., İlbaş, M., & Dönmez, G. (2013). Utilization of LPG and gasoline engine exhaust emissions by microalgae. *Journal of Hazardous Materials, 246, 247*, 173–180.

Taştan, B. E., Duygu, E., İlbaş, M., & Dönmez, G. (2016a). Enhancement of microalgal biomass production and dissolved inorganic C fixation from actual coal flue gas by exogenous salicylic acid and 1-triacontanol growth promoters. *Energy, 103*, 598–604.

Trozzi, C., Rentz, O., Oertel, D., Woodfield, M., & Stewart, R. (2010). Energy industries: Combustion in energy and transformation industries. *EMEP/EAA Air Pollutant Emission Inventory Guidebook*. European Environment Agency.

Van Den Hende, S., Vervaeren, H., & Boon, N. (2012). Flue gas compounds and microalgae: (Bio-)chemical interactions leading to biotechnological opportunities. *Biotechnology Advances, 30*, 1405–1424.

Van Den Hende, S., Vervaeren, H., Desmet, S., & Boon, N. (2011). Bioflocculation of microalgae and bacteria combined with flue gas to improve sewage treatment. *New Biotechnology, 29*, 23–31.

Van Thriel, C., Schäper, M., Kleinbeck, S., et al. (2010). Sensory and pulmonary effects of acute exposure to sulfur dioxide (SO_2). *Toxicology Letters, 196*, 42–50.
Vassy, A. (1965). Atmospheric ozone. *Advances in Geophysics, 11*, 115–173.
Wang, B., Li, Y., Wu, N., & Lan, C. Q. (2008). CO_2 bio-mitigation using microalgae. *Applied Microbiology and Biotechnology, 79*, 707–718.
Weil, E. D., & Sandler, S. R. (1997). Sulfur compounds. In *Kirk Othmer Encyclopedia of Chemical Technology* (pp. 275–340). New York: John Wiley and Sons.
WHO (World Health Organization). (2000). *Air Quality Guidelines* (Chapter 5.5: Carbon monoxide). Copenhagen, Denmark: WHO Regional Office for Europe. https://www.euro.who.int/__data/assets/pdf_file/0020/123059/AQG2ndEd_5_5carbonmonoxide.pdf (accessed on 26th February 2026).
Yoshihara, K. I., Nagase, H., Eguchi, K., Hirata, K., & Miyamoto, K. (1996). Biological elimination of nitric oxide and carbon dioxide from flue gas by marine microalga NOA-113 cultivated in a long tubular photobioreactor. *Journal of Fermentation and Bioengineering, 82*, 351–354.
Zeng, X., Danquah, M. K., Zhang, S., et al. (2012). Autotrophic cultivation of *Spirulina platensis* for CO_2 fixation and phycocyanin production. *Chemical Engineering Journal, 183*, 192–197.

Index

D

J

K

L

M

N

T

U